What Is Design Today?

what

is des

today

George H. Marcus

Harry N. Abrams, Inc., Publishers, New York

Commissioning Editor: Eric Himmel
Editor: Richard Olsen
Designer: Opto Design
Production: Maria Pia Gramaglia

This publication coincides with the
exhibition *What Is Design Today?*
organized by the Design Center
at Philadelphia University, supported
by the Philadelphia Exhibitions
Initiative (PEI), and funded by
The Pew Charitable Trusts.

Published in 2002 by Harry
N. Abrams, Incorporated,
New York. All rights
reserved. No part of the
contents of this book may
be reproduced without
the written permission of
the publisher

Printed and bound in Spain
10 9 8 7 6 5 4 3 2 1

Harry N. Abrams, Inc.
100 Fifth Avenue
New York, N.Y. 10011
www.abramsbooks.com

LIBRARY OF CONGRESS
CATALOGING-IN-PUBLICATION
DATA

Marcus, George H.
  What is design today? /
George H. Marcus.
  p. cm.
Includes bibliographical
references and index.
  ISBN 0-8109-9081-4 (pbk.)
  1. Design. I. Title.
  NK1510 .M37 2002
  745.4--dc21

2002006576

Abrams is a subsidiary of

 LA MARTINIÈRE
GROUPE

Acknowledgment is made
to the designers and
manufacturers who have
supplied photographs of the
products illustrated and to
the following photographers:
Will Brown (1, 10, 36, 49,
50, 77, 86, 96, 107, 108);
Urban Hedlund (66); Steve
Labuzetta (18); Marsel
Loermans (26, 87); Dean
Powell (84); Marino
Ramazzotti (70); Marcus
Rose (60); Camilla Sjodin
(95); Hans van der Mars
(69, 71); Bruce Weaver (72);
and Graydon Wood (3, 4).

Contents

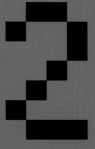
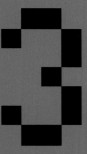

Introduction: What Is Design Today?

1
Since 1985, the annual volumes of the *International Design Yearbook,* compiled by Laurence King Publishing in London, have documented the yearly crop of consumer products, selected and introduced by noted designers and architects.

If you are reading this, you probably have some idea what design is—we all sort of do, we just can't quite put a finger on it. Dictionaries are no help. They define "design" as a sketch or a concept or a process but ignore the common usage that describes what these activities lead to. For those who understand the undefined meaning of this term, "design" signifies all of the objects that surround us—the clothes we wear, the products we use, the vehicles we ride in, the media that communicate with us graphically—which is what this book is about. The lack of a dictionary definition is perplexing since we are accustomed to seeing this term everywhere, on the cover of magazines, as the subject of exhibitions and books (figure 1), and as the focus of entire museum departments. The term "design" has been used in this way for more than a century, even before 1915, when Britain's Design and Industries Association was founded, or 1950, when the Museum of Modern Art in New York City and the Merchandise Mart in Chicago jointly initiated their influential Good Design exhibitions. As early as 1938, the British writer Anthony Bertram lamented the confusion over terminology in his book *Design,* based on a series of lectures on the BBC. "The very word 'design' is a mystery to the common man," he wrote, "almost a clique-word."[1] It still is.

While we all sort of know what design means, coming up with a definition that everyone agrees upon is not that easy, which is perhaps why the dictionaries have not tackled it. Is design an inclusive, value-free term applied across the board to describe all of the products of our day, or is it exclusive, with a particular bias about what an object should look like and what it should be? Must design be modern in its concept, or can it be postmodern—or, heaven

2

Florence Knoll's model living room designed for "An Exhibition for Modern Living" at the Detroit Institute of Arts in 1949 demonstrates the principles of the Good Design aesthetic: simplicity, utility, and economy; emphasis on natural materials; and disdain for applied ornament. (Photograph © 1949 The Detroit Institute of Arts)

forbid, traditional? Does design have to be produced in a factory, or can a single, lovingly handcrafted object also be design? And must it be useful, or do ornamental objects belong to design as well? Finally, how do such issues as accessibility, ecology, and appropriate technology fit in? Where, in fact, does design begin and where does it end?

The reason we even ask these questions is that we are still wrestling with the legacy of Good Design, a mid-twentieth-century populist movement that attempted to bring products with an economical, no-nonsense, modernist aesthetic to ordinary households (figure 2). Good Design was based on ideas that had originated in mid-nineteenth-century England, when reformers advocated a simple, utilitarian approach to the creation of everyday products as an alternative to the mass of elaborately decorated manufactured goods that the Industrial Revolution had made possible. Instead of accepting the furnishings in a variety of historic and decorative styles that were typically used in their time, Bertram and other proselytizers of the Good Design aesthetic on both sides of the Atlantic sought to impose their own preference for modern forms characterized by austere simplicity, natural materials, and the absence of ornament.

In a book entitled *What Is Modern Design?* published in 1950, the Museum of Modern Art explained the principles of Good Design to postwar consumers. Although the title was posed as a question, the text by the museum's curator, Edgar Kaufmann, Jr., made it very clear that there could be no question about the answer, and he proceeded to list twelve precepts that explained it all:

1. Modern design should fulfill the practical needs of modern life.
2. Modern design should express the spirit of our times.
3. Modern design should benefit by contemporary advances in the fine arts and pure sciences.
4. Modern design should take advantage of new materials and techniques and develop familiar ones.
5. Modern design should develop the forms, textures, and colors that spring from the direct fulfillment of requirements in appropriate materials and techniques.
6. Modern design should express the purpose of an object, never making it seem to be what it is not.
7. Modern design should express the qualities and beauties of the materials used, never making the materials seem to be what they are not.
8. Modern design should express the methods used to make an object, not disguising mass production as handicraft or simulating a technique not used.
9. Modern design should blend the expression of utility, materials, and process into a visually satisfactory whole.
10. Modern design should be simple, its structure evident in its appearance, avoiding extraneous enrichment.
11. Modern design should master the machine for the service of man.
12. Modern design should serve as wide a public as possible, considering modest needs and limited costs no less challenging than the requirements of pomp and luxury.[2]

Good Design was far-reaching but short-lived in its validation of a modern style for the postwar generation as consumers from diverse economic and social levels embraced this new look for a limited time during the early 1950s. But its impact on the design community lasted much longer.

3
Outrageously individ-
ualistic works such
as Matteo Thun's
brightly patterned
Nefertiti ceramic tea
set, a product of
the Italian Memphis
group (1981), ques-
tioned the values of
Good Design and the
modernist aesthetic.
Subversively revisiting
the timeless geomet-
ric forms so loved by
modernism, Thun
exploited them for
superficial, decorative
effect. (Philadelphia
Museum of Art)

4
The painted decora-
tion and stepped
silhouette of Robert
Venturi's postmodern
Sheraton chair for
Knoll (1978–84) are
knowing references to
the furniture of this
English eighteenth-
century cabinetmaker
that remind us of the
communicative possi-
bilities that can be
found in the history of
design. (Philadelphia
Museum of Art)

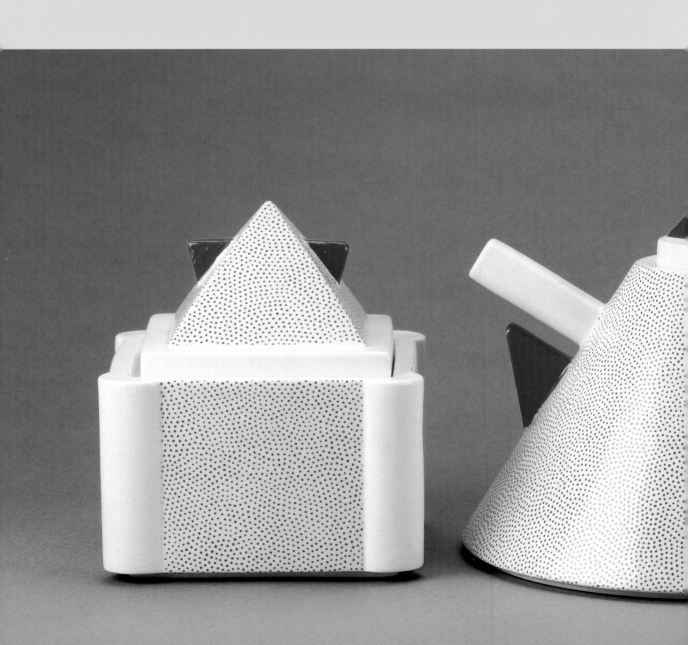

5, 6, 7

Design today encompasses a broad range of styles, from postmodernism and historicism to the most cutting edge of contemporary ideas. The subject of increasing public awareness, it has been promoted heavily through popular advertising, in specialized magazines, and on the web. Target's two-page advertisement introduced the postmodernist line that Michael Graves designed for this national retailer in 1999, branded with his "refined, sculptural, stunning" forms, signature colors, and signature, which appears on each of his products. Ethan Allen's furniture offers consumers both the comfort of tradition and a sense of contemporaneity through updated historic designs, such as these pared-down composites of eighteenth-century forms. Mossonline, the web site of the New York City store Moss, reflects its cool design principles, showcasing both modernist objects from decades ago and an array of more recent products, such as Konstantin Grcic's Mayday portable lamp for Flos (1998), shown at center, and Ross Lovegrove and Julian Brown's Basic thermos for Alfi (1991) in front of it.

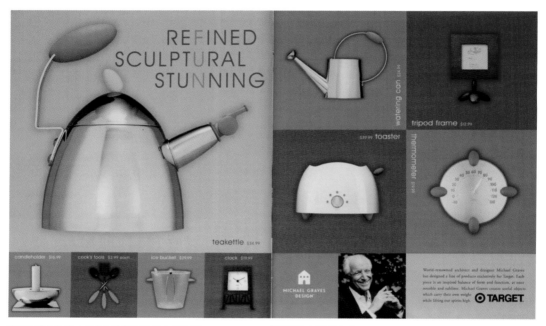

Not until the twentieth century was drawing toward its close did large numbers of critics, architects, and designers question the stylistic precepts of Good Design, asking why design had to be simple, when complexity was more interesting, or why it had to be clear, when obfuscation was more involving. Why not look to the past, they asked, not just the present, as inspiration for design? And what was wrong with ornament anyway, when it had served magnificently for so many millennia? As the tenets of dour postwar modernism loosened their hold, new models for design appeared, and we were introduced to an ever-expanding array of products with a new vocabulary of historical allusion, ornament, color, and form. This was spurred on by the outrageous, internationally celebrated Memphis group in Milan, whose colorful, densely patterned, and bizarrely shaped objects turned the tide of design in the early 1980s **(figure 3)**, and by the contemporaneous architect-led postmodernist movement, which was narrower in its gleanings, and far more intellectual, favoring allusions to classical and Western design history **(figure 4)**.

Now, half a century after the publication of *What Is Modern Design?*, we might revisit the propositions that were listed there so dogmatically and consider how these ideas relate to a definition of design for our own day. This time we cannot be so confident, since we do not have twelve easy precepts that will cover it all. Many "shoulds" have been removed from our vocabulary, and the principles and goals of design, which had been so clearly and tightly focused upon in Kaufmann's list, have expanded widely. Whereas Good Design adhered to a single aesthetic viewpoint, today the lines are not so distinct and we have no choice but to consider all stylistic possibilities. We must recognize, for example, that the postmodernism that infuses the products

Michael Graves designs for Target **(figure 5)** and the traditional styles that appear in the advertisements of Ethan Allen **(figure 6)** form a single continuum with the modernist and contemporary works favored by Moss, a specialty store in New York City **(figure 7)**. And whereas Good Design judged objects separately in terms of form, function, and economics, we must now judge them in a larger context that delves into the circumstances of their creation, production, and marketing and examine as well their social, environmental, and technological issues. Only in this way can we find satisfactory answers to the question that this book sets out to ask, "What Is Design Today?"

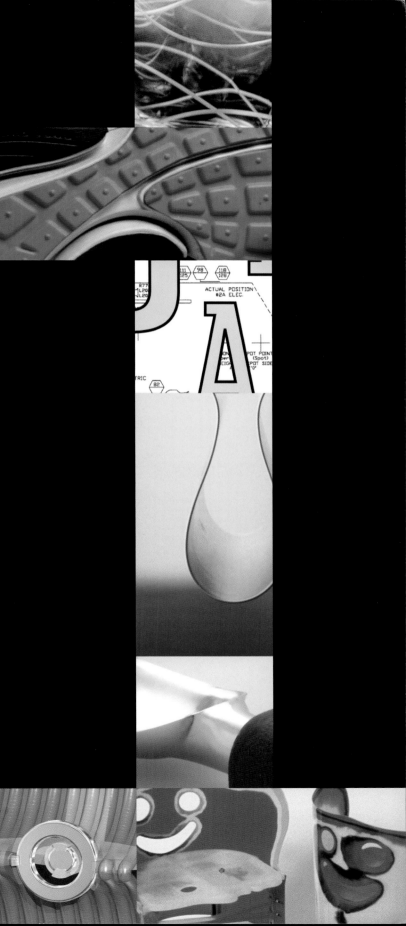

Understanding Process

In order to bring its Venus razor for women (**figure 9**) to the market in 2001, Gillette undertook a massive research and development campaign involving in-house design teams in the United States and Great Britain.

Included among its team of designers were (from left) Jill Shurtleff, who was responsible for the appearance and the ergonomic design of the razor, the sealed individual dispenser, and the in-shower storage unit; and Brian Oldroyd, who (with Frank Brown, not shown) developed the cartridge concept. Gerald Swanson did the engineering design of the cartridge;

Domenic Apprille, the engineering design of the handle; Charles Worrick, the engineering design of the in-shower unit; and John Petricca (not shown), the engineering design of the dispenser. Some fifty design and utility patents apply to this product.

Every object we see around us has been designed, and someone—a designer or group of designers—had to decide how each one would look, how each one would work, and how each one would be manufactured. Creating a product is most commonly the job of designers in the in-house design department of a corporation (**figures 8, 9**) or in an industrial-design firm (**figures 10a-i**); they work as a team and consult with specialists from other disciplines, such as management professionals, engineers, researchers, and marketing experts, on all aspects of its design. Theirs can be a massive collaborative effort, with many people making design decisions; faced with the realities of economics and the needs of the marketplace perceived by managers who may have limited experience with design itself, designers often find that they must defend the values of aesthetics and functionality that they are trained to bring to the design process.[3] Design also can be the province of individual firms, who think up ideas for new products and then convince manufacturers to produce and market them, or who receive commissions from manufacturers to develop new products to add to their lines (**figures 11, 12**). But some individuals, from inventors to fashion designers, also go it alone, manufacturing and distributing products themselves (**figures 13, 14**). Similarly, graphic designers work in small or large offices to give form to the books, magazines, web sites, advertisements, and other graphic messages that their clients want to communicate through print and digital media (**figures 15, 16**). Craftspeople usually work independently; they conceive an object and then, using specialized artisanal or technological skills, fabricate it by themselves or with a few assistants in a small studio (**figures 17, 18**).

The design and manufacture of a product can be a lengthy procedure, which begins when an idea comes to a designer or manufacturer and ends when the

9

Taking heed of what it learned from consumer surveys, Gillette created the Venus razor (2001) to meet women's shaving needs. The product includes triple blades, a shapely, ergonomically designed handle with waved ridges for easy gripping, and an in-shower storage unit with a holder for replacement blades.

10 a–i

How an Ice Cream Scoop Was Designed. In mid-2000, Bressler Group, a product development firm in Philadelphia, was asked by one of the nation's leading cutlery firms, Cutco, to help expand their product line into a wide range of kitchen utensils. From a large wish list of possibilities, Bressler Group initially created four products, a cheese knife, garlic press, can opener, and ice cream scoop, all with a family resemblance. To begin the research process, they bought a group of utensils made by other manufacturers from a kitchen shop, testing them informally and taking them apart to study how they were made. They also asked consumers to test these products and to offer feedback on what they liked and what they didn't like, videotaping the process so that they could review the reactions at a later date. Using the results of consumer testing, ergonomic standards from charts and books, and their initial sketches (a),

Bressler Group established their requirements for the products and decided the direction in which they should go to satisfy them. Since they were working on several products, Bressler Group decided to develop a handle style that all the utensils could share. The discovery that using the cheese knife, ice cream scoop, garlic press, and can opener required different grips and motions, however, resulted in slightly different final forms. During the concept stage, they began by experimenting with versions of the renowned Cutco handle, a scientifically based, ergonomic design, which had been developed by Thomas B. Lamb in the early 1940s, patented in 1945, and introduced by Cutco in 1952. But they also extended their investigation into other possible forms, carving many mock-ups of their own designs out of low-density foam (b).

Handles that had strongly defined bumps and indentations were rejected because they required specific placement of the hands, which meant that they would not have universal application and thus their comfortable use would be limited to certain segments of the population. At the same time as they were developing the handle, they were working out the designs of the individual products, deciding, for example, how the handle of the ice cream scoop would relate to the scoop head, which was being designed separately by the in-house design team of Cutco in Olean, New York, three hundred miles away. They used pieces of plain wooden dowel attached with a wing nut to experiment with the angle at which to attach the scoop (c). Next, they made a number of mock-ups with scoop heads attached and with slightly different designs to refine their concept (d).

In the design stage, after deciding in consultation with Cutco what direction the handle form should take, Bressler Group made a final mock-up of carved, painted, high-density foam (e), which included a stand at the end of the handle to keep the scoop from messing up the counter or tabletop; this became an important feature that would distinguish Cutco's product from those of its competitors. At the same time Cutco was refining its scoop head design, which changed from the flat head that Bressler Group had been using as a model, to a head that came to a point. Methods for construction of the utensil were investigated in drawings (f) and modeled in three-dimensional simulation as the design was transferred into electronic form via SolidWorks, a computer-aided design program. Now the construction of the ice cream scoop could be shown in exploded views (g)

and the product itself could be seen from different angles. The electronic file of the final design was then sent to Cutco, who combined it with their file for the scoop head and, using a computer-aided manufacturing program and prototyping machine, fabricated final prototypes. After testing the prototypes further with customer groups, Cutco sent the file to a factory in Taiwan for production. Bressler Group had presented their recommendations for overmolding Santoprene, a thermoplastic rubber material, and the level of hardness the Santoprene should have, along with the appearance design showing the handle in deep blue (h). While their material recommendations were accepted, at the last minute Cutco decided to manufacture the scoop in black with a partially textured grip, and the product was introduced to the market in this form early in 2002 (i).

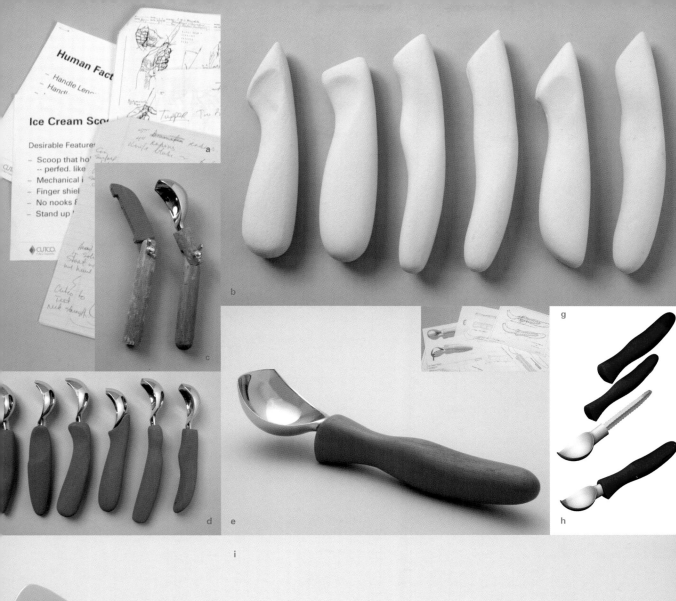

a

b

c

d

e

g

h

i

11
Karim Rashid, an Egyptian-born Canadian with his own studio in New York City, is one of today's most lionized designers, although it took many years and much personal salesmanship and effort to achieve this position. His work embraces virtually all fields, from the Millennium manhole covers he created for Con Edison in New York City to the furniture, plastic accessories, packaging, and lighting he designs for many different manufacturers.

12
Karim Rashid's tinted Murano-glass hanging lamp belongs to his series of Soft designs for George Kovacs (1999), which reflect the "sensual minimalism" of his work in all fields.

product is shipped from the factory. We can divide the process of design into four stages—research, concept, design, and production—although the design process is messy and these stages are never so clearly demarcated and frequently overlap.[4] During research, user needs are studied in depth. The design team often does market research, using focus groups and consumer testing; compares the function and construction of similar products by other manufacturers; and delves into human factors, including engineering and ergonomics, the study of the most comfortable, safe, and efficient ways that people and objects can interact. Through these research activities the criteria for the product's design are established (figure 10a). The concept stage uses sketches, study models (figures 10b, c), and preliminary engineering to formulate ideas for the product, answering the criteria questions established during the research process, and offers possibilities for the direction in which the product will go (figure 10d). The design stage gives physical form to the product, deciding how it is to look, how it is to be made, and what it is to be made of, and tests how it functions and interacts with the user. Designers today frequently use computer-aided design (CAD) and computer-aided manufacturing (CAM) software to expedite the design of the product and the production of drawings, models, and prototypes. Working models (figure 10e) and models that represent a product's appearance are constructed and the methods of manufacture and assembly are resolved (figures 10f, g). The design is reviewed for safety and other regulatory approvals, and final prototypes are built, tested, and evaluated once again. The fourth stage, production, includes making final appearance decisions (figure 10h), bidding the job out, selecting manufacturers, tooling up, manufacturing the product (figure 10i), and creating packaging.

13
James Dyson studied furniture and interior design but found his métier as an inventor. He built 5,127 prototypes of his bagless vacuum cleaner before he was satisfied with its design. Although the finished prototype was published immediately, he was unable to find investors in Europe to finance its production and went instead to Japan, where his invention achieved great success. Using the proceeds, he returned to England and in 1993 opened his own factory, producing what would become Europe's best-selling line of vacuum cleaners.

14
James Dyson's Root 8 Cyclone (2001) is the most powerful version of his series of bagless vacuum cleaners. Its futuristic, rocket-like form displays the distinctive yellow-and-gray palette that distinguished his first European model and established a product iconography that has been mimicked by Dyson's imitators. By dispensing with the vacuum cleaner bag, he was able to increase suction and keep the air clean from dust.

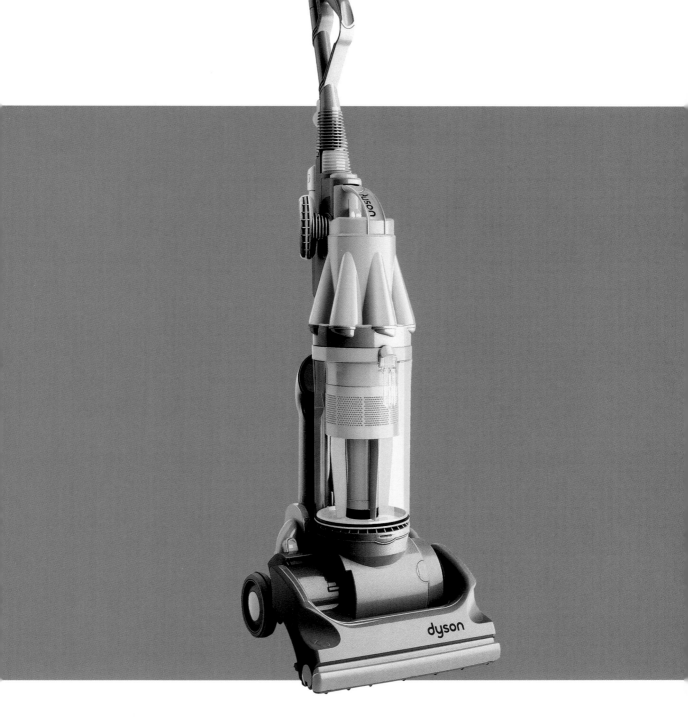

15
Although she studied art and aspired to be an illustrator, Paula Scher's career has been in advertising and graphic design. A highly influential designer, she worked independently and in a partnership with Koppel & Scher, but in 1990 she joined Pentagram, a large, international graphic-design firm. Scher is particularly known for her striking designs for cultural organizations.

16
Paula Scher's identity program for a new dance group, Ballet Tech (1996–97), combined her signature use of expressive typography with powerful photographs turned into design elements. The mirror imagery and doubled lettering set against a stage-lighting diagram reinforce the "tech" aspect of the company's name.

17

The craftsman Wendell Castle is an elder statesman of American woodworking. For some forty years he has been creating organically shaped furniture that ranges from utilitarian to purely sculptural pieces.

Castle studied sculpture, but he also received a degree in industrial design, and he has tried to balance his one-off studio pieces with works that would be available to a larger audience. He created a series of molded-plastic chairs during the late 1960s, has issued limited-edition lines, and in 2000 began to market his own collection of production furniture.

18

Wendell Castle's Last Chance (2001) combines a copper-clad cabinet that has several drawers set in at odd angles with a Douglas fir table, alluding to a functional purpose that is, however, no longer necessary for a definition of crafts.

By parsing the stages of design as we have done, it can be seen that some of the activities undertaken by the industrial designer are the same as those of the craftsperson, although the goal of the former is usually reproduction in numerous identical examples and of the latter, limited fabrication in one or a few copies. The making of models, molds, and prototypes by the industrial designer follows many of the same processes that the craftsperson uses. The two disciplines come most closely together at this point in the process through the similarly exact working of materials, regardless of whether it is done by hand, machine, or computer, to create a single example that reflects all aspects of the creative effort. What is done with that single example, whether it stands alone to be used and admired or serves as the prototype for many replicas, generally distinguishes one approach from the other. But some of the widely published examples of recent design have been made only as prototypes or in one-off or small-batch production **(figure 19)**, and some of the most technologically sophisticated creations of contemporary design—space vehicles, for example—also are made this way, the accomplishment of extremely refined, one-of-a-kind workmanship. Alternatively, some examples of craft production appear in large series, such as the hand-blown glass vases and other objects designed by Simon Pearce and made in his workshops in Vermont and West Virginia.

The term "crafts," as we often use it, does not define the product but the intent and self-categorization of the maker. Being a studio craftsperson in today's postindustrial society can be a political act, a decision to maintain a high level of quality through personal production, or a decision to remain small, elite, and often self-consciously "artistic." This follows the trend that crafts have taken since the 1960s, rejecting the primacy of utility in favor of self-expression and aspirations to the world of the fine arts, although the contributions of an extensive group of craftspeople who remain devoted to creating functional objects recently have been brought into focus.[5] Crafts have no locus in form, process, or material: They are often very much in sympathy with the cutting-edge aesthetic, the technologies, and the computer-aided development of objects of industrial design, and today, crafts do not even have to be made by hand. Stanley Lechtzin, a noted American metalworker, no longer executes his own designs directly in a tangible medium; he creates his jewelry on a computer and relies on the digital capabilities of Rapid Prototyping to manufacture the objects for him **(figure 20)**. "I now believe that the CAD/CAM is a new craft medium," he says, finding support in the arguments of Malcolm McCullough's *Abstracting Craft*[6] as he praises the advantages of working in this way. "This new medium brings together and places in the 'hands' of the artist, the three necessary stages of object making: concept, design, and execution....I know that when I am seated at the computer, I interact with the objects that I build in a manner that is an echo of the experiences that I had when seated at the bench...only it is better....I create elements and join them to one another. Rather than a torch and solder, I use Boolean unions. When I needed to get a better or more detailed view of my work I used to put on a jeweler's loupe. Now all that I need do is zoom in. I have precision of movement that I could not have dreamed of twenty years ago."[7]

Conventional distinctions between craft and design are losing their meaning. Craft today is no longer synonymous with a definition that requires an object

19
Not all design
is manufactured in
large quantities
by machine. Mind
the Gap (1998),
a rubber-and-steel
coffee table with a
magazine holder in its
center, is made in
small batches as
required. It was creat-
ed by El Ultimo Grito,
a design firm organ-
ized in 1997 by three
Spaniards (Roberto
Feo, Rosario Hurtado,
and Francisco Santos)
living in London.

20

Stanley Lechtzin designs his jewelry on a computer and produces it with the Rapid Prototyping process, the method by which his RarEarth 2 brooch of nylon, epoxy, and rare-earth magnets (2001) was created.

During the creative process he can easily view his works from different angles; his RarEarth 2 brooch is seen here in computer simulation from top, back, and right and in perspective. After he is satisfied with his designs, they are manufactured directly by Rapid Prototyping without his further intervention, an example of a craftsman adapting his approach to today's latest technologies.

Top

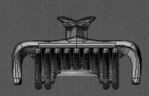

Perspective

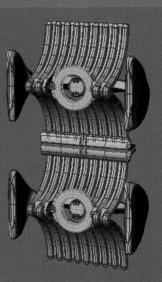

Back

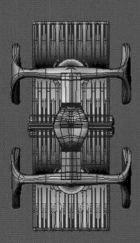

Right

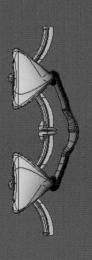

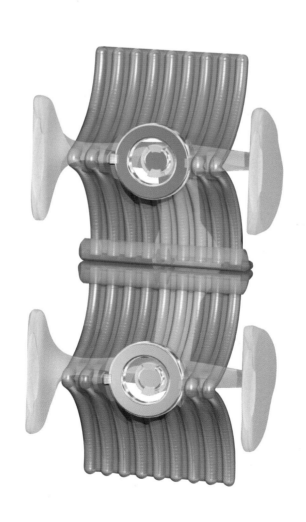

21
Gaetano Pesce
creates his Open Sky
resin furniture and
lighting in a half mold
"open to the sky"—
the derivation of his
company's name. His
Baby Crosby child's
chairs (1999) are made
by this method in
series; each is unique,
their variances due
to the hand of the
worker, the color of
the resin, and its
transparency.

to reveal its workmanship nor one of industrial production that demands that goods be replicated exactly in large numbers by workers on an assembly line, each turning the same screw or assembling the same part repeatedly all day, much as Charlie Chaplin did so poignantly in *Modern Times.* Many other production scenarios are now possible. In the factory-made objects in his "diversified series production," the Italian designer Gaetano Pesce tries to introduce variations so that each example appears individually made and unique **(figure 21)**. "In the future," he has said, "customers will expect original objects. What I call the third industrial revolution will give people the opportunity to have a unique piece; the technology we have today gives us the possibility to produce in this way. Materials too. It is very much like what the artisans of the past achieved. But at the same time, it reflects the spirit of our time, where everything is relative."[8] Large manufacturers also have begun to introduce variance during industrial production. The computers that now contribute to all aspects of design and manufacturing and assure the regularity of factory-made goods also can be used to do just the opposite, to introduce variations throughout a production run. This was the goal set by Karim Rashid for a series of tabletop products for Nambé. Using random software programs, he was able to assure that his glass and metal vases would be individualized so that consumers looking at a display would be able to choose from a number of different examples and feel they were buying an object that was unique **(figure 22)**.

Individualization and customization of goods are the directions of much of postindustrial manufacture, and their impact is certain to increase as computers take over many aspects of production. Customization serves niche markets with products that, to counteract homogenized globalization, have been adjusted to meet the specific functional needs or cultural values of targeted groups, a process that was pioneered on a worldwide scale by Philips during the 1980s. It was also the secret behind the success of Swatch, the Swiss company that made watches into fashion accessories by constantly creating new collections with a large variety of styles and colors to make them seem fully individualized to the consumer. This new direction, "the short run of partially or completely customized products," emerged in the 1960s, and it was described in 1980 by Alvin Toffler in his book *The Third Wave.*[9] He showed how many firms throughout the world—from those such as Hewlett-Packard, which assembled electronic equipment to order from a mass of ready components, to garment manufacturers and T-shirt printers, who could quickly adjust the sizes, colors, and designs of their products for different locales and consumer groups—were using advanced computer technologies to customize their products. In doing so they were able to limit their investment in an inventory that might not be sold. More recently, what is known as "mass customization"—configuring unique products on a greatly enlarged scale for smaller markets and individuals—has become the watchword for big business and the advertising industry.[10] With the aid of CAD/CAM, even textile mills, with their large and costly equipment, can now turn out short runs of customized fabric efficiently and economically. Mass customization gives businesses the economic advantage of flexibility to update products according to demand and incorporate new technologies very quickly. They are able to fit products to changing styles and needs of individuals and, at the same time, charge premium prices for products that are customized to meet those needs. Now, as Toffler predicted, we are coming

Karim Rashid took
advantage of a fully
automated process
to create variance
during the production
of his Motus collec-
tion of lead-crystal
objects for Nambé;
the cut lines, he
explains, were
"designed with a
digital tool path so
that the cuts can
change randomly
from object to object
and always meet up
around changing
circumferences. This
is a mass-produced
way to create non-
serialized objects."

where the customer is "so integrated into the production process that we will find it more and more difficult to tell just who is actually the consumer and who the producer."[11]

Customizing of course is not new. Well-off purchasers have always ordered custom products, particularly made-to-measure clothing, directly from their suppliers to assure the quality and exclusivity of their wardrobes. Ordinary Americans also have been able to customize some of their purchases, notably when they go out to buy a new car. This practice reached extremes during the 1950s, when the need to sell large numbers of models annually caused fierce competition among American automobile manufacturers. Customers were enticed with the possibility of personalizing their cars with individual color combinations; in 1954, for example, Chrysler offered a "record choice of 58 exterior colors and 86 two-tone color combinations,"[12] along with a large array of choices for the interior furnishings. Other types of customization also have been available: The introduction of modular furniture elements early in the twentieth century allowed consumers to individualize the arrangement of home storage while the component method for creating stereo systems also has been favored by audiophiles for a long time.

Today customization is at hand for all of us on the World Wide Web as we instruct our servers to create home pages that automatically give us our horoscopes, follow our stocks, and report the scores of our favorite teams. Manufacturers also are making customization available directly to consumers in the same way, and we can now buy customized clothing through their web sites. We can, for example, modify the elements of Levi's jeans or even design our own jeans from scratch with their help,[13] and we have the option of sitting at home and designing many more of our products. With real-time communication available, the focus of mass customization has shifted from the efficiency and economy of the manufacturer to the emotional satisfaction of the consumer, much as Toffler predicted. As NikeBiz's question-and-answer web page put it, the Nike iD sport-shoe line introduced in 1999 "is a totally new approach in product design, which allows consumers to literally leave their mark on the shoes they wear. With Nike iD, consumers can express themselves on their footwear. Each shoe is 'made-to-order,' combining self-expression and product preferences with Nike's expertise in creating performance products. With Nike iD, you don't have to be Michael Jordan or Mia Hamm to have your own signature shoes."[14] Nike's web site guides us through a series of steps in which we can select a pattern if we like, choose base and accent colors from a chart, and then add our own identifying signature, or "string" **(figure 23)**, viewing the customized product from six different angles at every stage of its transformation **(figure 24)**. After we approve our designs, our shoes are made to order and shipped to arrive at our doors within two to three weeks. Many other businesses offer similar services, from building our own Dell computer[15] to individualizing our own factory-made house[16] to formulating our own scent.[17]

The success of this revolution in production can be attributed to the increasing use of CAD/CAM, but it also has been aided by the introduction of a manufacturing system that separates the fabrication of components, often made in many different factories in different parts of the world, from the assembly or finishing of final products. Much of this is now done by hand in

23

This pair of customized Air Presto sport shoes was ordered by the author on the Nike web site and was received at his home in Philadelphia three weeks and two days later.

The shoes were made in Korea and arrived at Nike headquarters, in Portland, Oregon, sixteen days after the order was placed. The personal string "Design," chosen by the author, is incised into the plastic lacing. The one-piece cardboard shipping carton, which is printed on the interior and uses no harmful glues or inks in its fabrication, includes instructions for flattening it out to make a poster or for refolding it to make a pedestal for displaying the shoes.

24
As we follow the procedure for customizing on Nike's web site, we can choose to see the shoe from six different angles at each stage as the design evolves. Nike made production and customization more efficient and economical by creating a shoe with a stretch mesh upper that could fit a range of foot sizes, replacing the conventional numerical sizes with generic sizes, from XXS to XL.

third-world countries, where labor is cheaper. Products assembled from components abroad with lower labor costs can be individualized to order quickly and economically. The move to offshore production has been led by multinational corporations, whose brand-identity programs demand large-scale manufacturing but who have few or no factories of their own. These large businesses are accustomed to contracting out the production of goods to manufacturers who operate their own factories in free-trade zones abroad (or subcontract the work to others), allowing the corporate offices to concentrate instead on branding, which is creating an image that goes beyond marketing and identity to build up brand loyalty based on the promotion of a particular lifestyle. The existence of these often-complicated production arrangements is verified by the indications of origin on many of our everyday products. The expansion of this production model, with its widely reported pattern of corporate abnegation of responsibility, ecological indifference, and concomitant questions of human rights violations toward the indigenous underemployed populations recruited to manufacture these goods,[18] has led to considerable controversy and a new international activism, much of it centered on college campuses. One result has been that some universities have instituted strict manufacturing codes for objects that carry their names and logos,[19] influenced by the standard for workplace conditions established by Social Accountability International.[20] But the almost universal market for electronics and the products of fashion and style keeps the system going, with the availability of less expensive merchandise taking precedence over environmental and humanitarian concerns for many consumers.

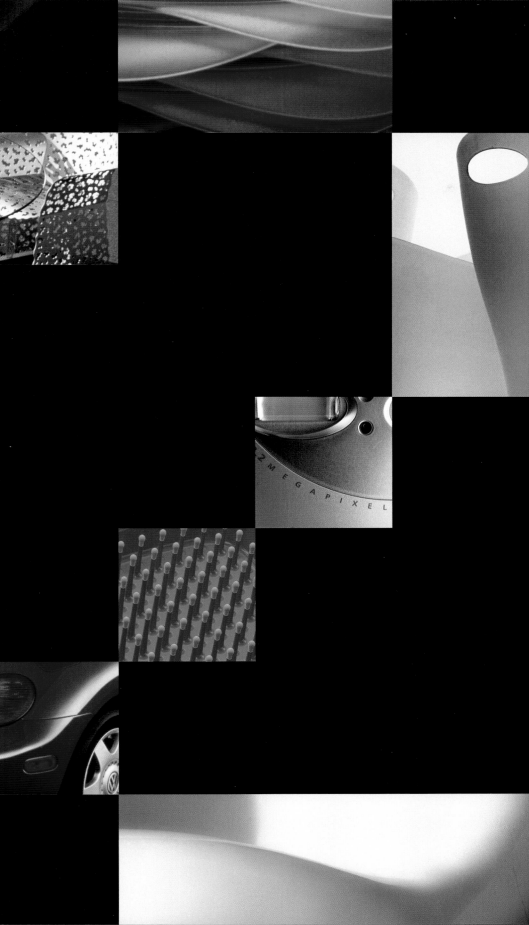

le

Style is the aesthetic wrapping in which all of our products come, but they don't all come wrapped in the same style. Just how the elements of design—shape, color, texture, material, pattern, and ornament—are put together, and the messages they are meant to convey, is what separates products from one another and defines their style. It is what makes objects desirable beyond their function and what creates the basis for personal preferences and taste. Style is both the individual expression of each designer's vital imagination, which means that designers' works are often distinctly recognizable as theirs alone, and the expression of a cultural moment set in the context of history, in which all the works of one era seem to be part of a general pattern of their time. We seem to recognize easily the characteristics that define the styles of earlier periods—the organic tendrils that dominate Art Nouveau, the angular, cubistic shapes of Art Deco—but not those of our own time, because we get caught up in the details of contemporary products and the work of contemporary designers, unable to decipher the unifying elements that delineate the style of today.

We live in an eclectic age. At the same time as a ubiquitous stylistic conservatism brings corporate blandness to the majority of our electronics and appliances, be they black, white, or metallic boxes, we welcome new approaches that introduce bright colors, curvilinear forms, less valued materials, and narrative and pictorial content into our homes, allowing them to expand our stylistic outlook. We encourage designers to explore individuality and self-expression in giving objects unique style. We see its extreme in the idiosyncratic work of Ingo Maurer, who brings both wit and technological inventiveness **(figure 25; see also figure 67)** to the field of lighting, and in that of

Droog Design, a Dutch association of designers who share a sharp, dry *(Droog)* humor in their approach to products, favoring unpretentiousness as an antidote to the slick design aesthetic that has dominated recent times **(figure 26; see also figures 69, 71, 87)**. But nothing could be more representative of today's permissiveness than the array of outfits parading the fashion runways, where along with ever-new approaches to draping fabrics and constructing garments, exotic and historic samplings mingle with riffs on popular culture to create fusions of the richest invention **(figure 27)**. A similar approach to the mixing of stylistic elements is being taken in interior design, although it is generally not so flamboyant, where antiques and period-style furnishings (including such twentieth-century reproductions and revivals as the metal furniture of the twenties and the newly popular Stickley lines) coexist with examples of recent creation. This freedom to borrow from a pool of available ideas is the legacy of the postmodernist thinking of the last two decades, which, in broadly rejecting the objective, formalist vocabulary of Good Design, opened up design to stylistic influences from all eras and all areas of the world and brought a greater complexity to our surroundings.

Referencing today comes in many formats; it has no manifesto, and it is all over the map, reclaiming everything from vernacular clichés **(figure 28)** to retro-futuristic fantasies **(figure 29)**. It may be expressed through pattern and representation, as is the case in the abstract, openwork decoration of Richard Schultz's shrub-like Topiary garden chairs based on cloud forms and floral elements **(figure 30)** and in the realistic bucolic imagery on Constantin Boym's

28
Marcel Wanders coyly chose the Jack-be-nimble candlestick as the model for his steel B.L.O. lamp for Flos (2001). He carries the storybook reference to its unexpected but logical end by incorporating sensors that allow us to extinguish the electric, flame-shaped bulb by blowing on it, just as we would with a candle.

29
Making reference to today's strong interest in mid-century modern design, the Italian appliance manufacturer SMEG created this refrigerator with a thirties, streamlined form and a typical fifties color (2001).

Upstate dinnerware (figure 31). It is seen as well in the vibrant introduction of narrative and figural content into such products as Gaetano Pesce's Baby Crosby chairs with their bright, smiling faces (see figure 21) and the ever-growing family of "cutensils," humanoid kitchen and bath implements originally conceived by Orange and manufactured by Koziol (see figure 102). It also may be manifested through the manipulation of historic forms, as in the eclectic furniture in Ethan Allen's line (see figure 6) and the redesigns and updates of Scandinavian mid-century modern furnishings sold by IKEA, the international retailer from Sweden. But referencing also can be very subtle in its allusions, for example, the warm paint tones drawn from the colors of the eggshells of Araucana hens that for Martha Stewart suggest the ambience of faded Victorian Americana. With Stewart's immense audience and impact, this palette became widely popular after it was introduced in 1995, not only influencing interior decoration and home furnishings but also crossing over to impact on the colors favored by graphic designers.

The international automobile industry also has been a hotbed of historical revival, with its evocations of popular models from the past, among them, Volkswagen's new Beetle, the Ford Thunderbird, and the Mini Cooper. Aside from the overall reference to the shape of its predecessor, the Beetle (figure 32), however, has little to do with the small, cheap, pared-down, "ugly" car that enjoyed wide popularity in the United States during the 1950s and 1960s. With its bright crayon colors, the new Beetle was conceived as a toy for Americans of the nineties, a little brother to its elegant, allied model the Audi TT, sharing its ample, rounded, structurally secure lines and its overall look of contemporaneity. Attempts at engendering nostalgic reactions, such

30
Richard Schultz's Topiary outdoor furniture (1996), inspired by the patterns of dappled light filtering through the shrubbery in his garden, refers to the natural surroundings for which it was designed. The shape is meant to suggest the clipped shrubs that give this collection its name while floral and cloud-like forms inspired the openwork decoration in the bent aluminum material.

31
Like a number of other cutting-edge designers, Constantin Boym plays with digital photography, introducing its images into his ceramic Upstate dinnerware for Handy (2002) as he contributes to the reemergence of narrative design today.

32
The Volkswagen Beetle became the best-selling automobile revival in the 1990s, with its nod to nostalgia tempered by a dose of modernity and its earlier restraint supplanted by sophisticated styling, humor, and eye-catching colors.

33
The PT Cruiser (2000), Daimler-Chrysler's bid for a competitive revivalist model, is not based on any specific 1930s automobile but makes a generalized nostalgia its selling point. While it tries not to look contemporary, its regulation bumpers and the comfort and conveniences of the interior bespeak its moment in design today.

as the "Flower Power" vase that comes attached to the Beetle's dashboard, make the car just that much more endearing to superannuated flower children. But they, like drivers of a younger generation who may be clueless to the joke, now also prefer ease, comfort, and luxury in their cars. At the other end of the retro-vision spectrum, Daimler-Chrysler's equally successful PT Cruiser (figure 33) eschews contemporaneity in its concept and its styling as it offers a design with less humor and seemingly more cynicism. While exploiting a fascination for the opulent 1930s movies lifestyle, the Cruiser does not directly revive an earlier model but comes across as a generic, almost Disney-like, version of the past, with identities that change from viewpoint to viewpoint—a touring car from the front, a station wagon from the back, and a gangster's getaway car, with vestigial running board, from the side.

Perhaps the most up-front referencing with the most far-reaching possibilities is that of Tom Dixon and Jasper Morrison, two highly visible British designers who in Dixon's word seek to "rethink" ordinary objects as they return to a form of stylistic modernism characterized by extreme simplicity. Their approach has been to take specific, sometimes anonymous, products and reconsider them, how they are to be remade and how they are to be "reused," purifying their forms and taking them back to the essences of their designs (figures 34, 35). They both have shared their appreciation of the everyday in print. In his book *A World Without Words*,[21] Morrison compiled a set of hopeful images of objects from the past that he admired, much as Le Corbusier had done in *The Decorative Art of Today* (1925), where he sang the praises of such standardized products as office furniture and industrial glassware and promoted them as the basis for design in the future. Dixon's

35
The understated neomodernism of Jasper Morrison's Plan storage system for Cappellini (1999) exemplifies his back-to-basics approach as he references earlier sources from the history of design.

36
In order to stimulate the creation of new products for its style-conscious PS collection, IKEA encourages its designers to investigate new materials, techniques, and uses for established objects. Ehlén Johansson took a second look at the mundane white ceramic fuses that were being made in a factory in Poland and re-created them as linking tea-light holders without hiding their industrial origins.

37
Philippe Starck's
Bubble Club sofa and
armchair for Kartell
(2000) reconceptual-
izes a typical uphol-
stered model in a
completely new mate-
rial, polyethylene, but
retains the allusion to
its original format in
the phantom "cush-
ions" outlined in its
surface. Industrially
made from a weather-
resistant plastic, this
series brings a living-
room favorite out into
the open.

book *Rethink* takes a similar stand as it revisits contemporary design; it is
"about looking at the world of made objects in a different way. Trying to
find the hidden beauty in the mundane. Spotting fitness for purpose where it
exists, even when not intended. Sometimes taming the industrial artefact for
domestic use. But always keeping your eyes wide open."[22] Dixon looks at
industrial products and revisits them much like High-Tech designers
did in the 1970s, suggesting inventive ways and different situations for
reintroducing them intact into our landscape.[23] Other designers also have
been rethinking and purifying earlier forms in line with Dixon's call for
"taming" industrial artifacts. Ehlén Johansson has taken a simple ceramic
fuse, which she came upon when visiting a factory in Poland, and
reconfigured it into a decorative tabletop product for IKEA **(figure 36)**. Philippe
Starck, the French designer who since the 1980s has demonstrated the
marketing validity of a personal, exuberant, and, in his case, eccentric style,
more recently also has been reexamining and purifying; using new materials
and new processes, he has revisited the type objects that Le Corbusier
ennobled in his writings. His Bubble Club chair resembles an iconic version
of the upholstered form that Le Corbusier celebrated, but instead of the
traditional soft and comfortable handmade version, Starck's is formed
industrially of hard, molded polyethylene **(figure 37)**. This is one of many
products that are returning plastic to the position of value it enjoyed in the
1960s and 1970s. The type designer Zuzana Licko also is rethinking the past
and, in her case, her own earlier designs, restructuring the coarse, digital
typefaces she created in 1985 into the Lo-Res family of type in 2001 **(figure 38)**.
The renewed popularity of techno-graphic imagery and low-resolution
typography is reflected as well in the structural, grid-based lettering designed

Setting Style

Lo-Res Plus 9 Bold Wide   Copyright 2002 Emigre Inc.

38
In 1985 Zuzana Licko designed several coarse bit maped type fonts for low-resolution computer screens and printers, which she reconfigured into Émigré's Lo-Res type family in 2001. Once high-resolution technology arrived, these had little use, but they have now made a comeback, found frequently in graphic design applications that express subjects related to digital technology.

by John Klotnia for use as the display face in this book.

Without ignoring the trend of direct referencing, a separate impulse has emerged that sets fluid, organic, sculptural form as a dominating force in design today, a style defined by Karim Rashid as "blobism" **(figure 39)**. Rashid has achieved celebrity for the breathless beauty of the everyday objects he creates in these forms, depending on the sophistication of computer software to achieve them. "The world around us seems to be perpetually getting softer," he notes. "Our objects are softer, our cars are rounder, our computers are blobier, and even our bodies are fatter." But "Is Soft a style?," he asks. "Is it an axiom of our new physical landscape? Or is it a desire to conquer in all mathematical probabilities, forms that could previously never be documented in any rational formulae...a result of new computer-aided tools of morphing?"[24] Rashid is best known for the range of sensuous plastic products he created for the Canadian manufacturer Umbra. His molded-polypropylene Garbo trash can **(figure 40)**, with its elegant silhouette, mysterious translucent tonalities, and low prices that rack up truly universal sales figures, has become one of the icons of mass-market design today.

Approaching design as organic sculpture is not new and cannot be completely attributed to the advances of the computer despite the assertions of Rashid and other contemporary designers; elaborate organic forms defined the style of the 1950s, and even after they were repudiated, they continued as an undercurrent in the history of design, in products created for the space industry, in molded-plastic furnishings, in ergonomic sports equipment,

The complex curves in Karim Rashid's sensuous chairs for Magis (1999), which he labels "blobs," emerged from applying the capabilities of the computer as a design and engineering tool and demarcated a stylistic departure for the 1990s.

40
Karim Rashid's
high-impact, molded-
polypropylene Garbo
trash can for Umbra
(1996) became the
icon of commercial
success for its fluid,
organic style.

41
Kodak's Easy Share
digital camera (2001),
conceived with the
fluid, organic curves
that have been a part
of camera design
for the past decade,
sits like a piece of
sculpture in its dock.
Its base allows pic-
tures to be transferred
from the camera to
the computer.

42
Time magazine cele-
brated a new interna-
tional awareness of
the economic and
cultural advantages of
high style with its
cover story on design
in March of 2000,
presenting a plethora
of colorful plastic
products and high-
lighting the work of
the Australian design-
er Marc Newson.

as well as in the design of cameras, which established a type that remains evident today **(figure 41)**. The complex, sculptural style began to gain popularity again in the 1980s and 1990s, for example, with the organic Ford Taurus, introduced in 1986. Philippe Starck exploited it too in the furnishings of his famous hotel interiors and in many of his household products. But the greatest acclaim for design as sculptural form today has gone to the architect Frank Gehry, whose curvilinear, metal-clad museums and concert halls bring immediate celebrity to every locale in which they are built.

The sculptural approach was celebrated with a cover story in *Time* magazine in 2000 entitled "The Rebirth of Design" **(figure 42)**. *Time* clearly woke up to the economic advantages of emphasizing style in an era of prosperity and to the increasing popularity of such highly styled products on all levels of the economy when it announced, "From radios to cars to toothbrushes, America is bowled over by style." And the style they were bowled over by was organic, sculptural, and often very colorful. The designer to whom *Time* gave most coverage was Marc Newson, an Australian who works in London. He was singled out for the breadth of his work and especially for having been invited to design a concept car for Ford **(figure 43)**. Newson applies organic, rounded forms and a minimalist vocabulary to his products, bringing a stylistic unity to the vast disparity of the commissions he receives, whether for a polypropylene dish rack for Magis **(figure 44)** or an aluminum bicycle for Biomega **(figure 45)**.

In the late 1990s, sensuous forms combined with bright, translucent colors to become the most visible of stylistic markers. Every new product—from the most serious to the most frivolous—now seemed to come in shades we previously associated with the flavors of candy. It is no coincidence that when Apple was working on the translucent polycarbonate iMac computer in 1998, the design team sought out candy manufacturers for advice on how to keep color consistency within their product line. "We had to make sure that the color and level of translucency were exactly the same in the first computer and every one thereafter," explained Jonathan Ive, head of industrial design for Apple. "This led us to finding a partner who does a lot of work in the candy industry, because a lot of candies are translucent [and these] guys have so much experience in how you control the compounding and a great understanding of the science of color control."[25] Although Apple was not the first manufacturer to adopt translucent colors—fun products for the teen market such as telephones, pagers, watches, and portable video game players had used them from the mid-nineties—the company made news by going against the beige standard to add colors to serious, grown-up products. Starting out with "Bondi" blue, which alluded to an Australian surfing paradise to promote the iMac as Internet ready, Apple pushed the envelope further when it offered the iMac in blueberry, grape, tangerine **(figure 46)**, lime, and strawberry. It reinforced this conceit in its marketing, where these colors were advertised as "five fruit flavors" and the computers themselves as "iCandy" **(figure 47)**. The euphoria of five fabulously successful flavorful colors, and a translucency and fluidity that went along with them, soon reached epic proportions, bringing a quick dissolution into copycat marketing **(figure 48)**. But they also appeared in elegant, thoughtful new creations **(figure 49)** as well as in clever revisions of mundane products **(figure 50)**. Apple went on to refine its vision with products of an icy transparency that massed critical acclaim for

Featuring his signature bright-orange color, Marc Newson's blunt-ended concept car for Ford (2000) transgresses the codes of insider automobile design as it reveals a sense of humor, round and friendly organic lines, an accessible trunk that pulls out like a drawer, and doors that open from the center.

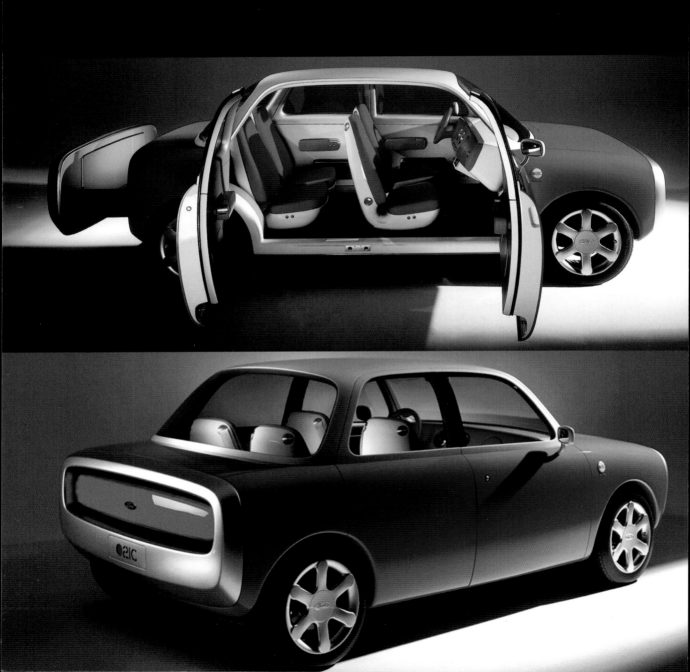

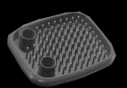

44
Marc Newson's
brilliantly colored
Dish Doctor dishrack
for Magis (1997)
turned an ordinary
object into a stylish
accessory, achieving
rapid celebrity as it
uplifted everyday
chores with its aes-
thetic validation.

45
The zigzag frame of
Marc Newson's urban
bicycle for Biomega
(1998–99) is made of
two pieces of vacuum-
formed aluminum
glued together using
airplane technology;
it hides the cables,
streamlines its appear-
ance, and glows in
the dark.

their design (figure 51), and, more recently, of milky white plastic, such as the iBook laptop computer and iPod M3 player released in 2001 and the new iMac, in 2002, but the colorful objects have evoked the broadest consumer response.

Like the Life Savers and Gummi Bears to which the translucent colors allude, these products were meant to be enjoyed and devoured. They suddenly influenced the look of our outfits, our homes, and our offices, where even workaday business machines have responded to this trend. But with these wonderful colors and their flavorful associations, the real question that these products asked was why not buy several? After having been raised amid prosperity and thoroughly pampered, today's generation naturally would expect to have more than one of everything—regardless of its cost, scale, status, or effect on the environment. "Having one [National Basketball Association] championship is like having one car," the Lakers' star player Shaquille O'Neal explained ingenuously. "It's not enough for me."[26] And one seems not to be enough for many others either, who have easily justified buying more and more products—for fashion as well as function—as technological breakthroughs, competition, and economy lowered their prices and as new features and applications were successively added.

In 1998 the Finnish telephone manufacturer Nokia introduced the Xpress-On interchangeable faceplates in different colors for its mobile phones and blatantly advertised it as the fashion statement that such devices had become (figure 52). The advertisement showed how you could easily coordinate your look just by changing Nokia's patented faceplates. The extraordinary success of

iCandy

Think different.

Cool
Colors,
Hot
Products

49
TDK's one-sided, petal-shaped CD-R shipping and storage cases of slightly flexible polystyrene (1999) can be stacked, carried individually, or hung on a hook and identified and organized by their rainbow colors. In creating this new CD packaging concept, IDEO, a leading international industrial-design firm, realized they could dispense with half the conventional packaging and half the material since CDs need to be protected on one side only.

50
These polypropylene ring binders from Storex Industries (1998), designed by Gad Shaanan Design, cleverly update a conventional product by playing with the colors and the sensuous forms of the new plastic revolution.

51
Hailed as one of the great stylistic achievements of recent communications design, Apple's translucent Cube computer (2000) had a commercial life of only about a year, after which style gave way to a more expandable model on the company's product list.

Nokia, the Finnish
firm that has led the
international craze for
cellular phones,
demonstrates that
keeping in touch with
friends and family is
more than just com-
munication—it is a
signifier of both style
and status.

Another Nokia discovery:

People like choices.

So we designed our Nokia
5100 Series digital phones
with Xpress-on™ covers.
These fashionable faceplates
snap on and off for a quick
change of color. Choose the
one that fits you.

NOKIA
CONNECTING PEOPLE

the Nokia line demonstrated that the ability to individualize and personalize was a stylistic trend that would sell products; soon the faceplates could be customized further by adding patterns and images while the possible rings one could access were legion. Other manufacturers in the world of electronics quickly copied Nokia's emphasis on the stylistic trappings of design. The marketing of common consumer products not for what they can do but for how they look demonstrates how much we have taken the advanced technology that surrounds us for granted. Again and again, as new product types become fully developed and integrated into our lives, we expect that they will all work equally well and we look for other clues to differentiate them and help us make our buying decisions. Competition demands that products be easily recognizable, and the creative use of form, material, and color is one way to achieve this. Aside from the inclusion of extra features and further miniaturization (especially desirable in portable devices), the only thing left to distinguish these new, widely available, complex products is their style.

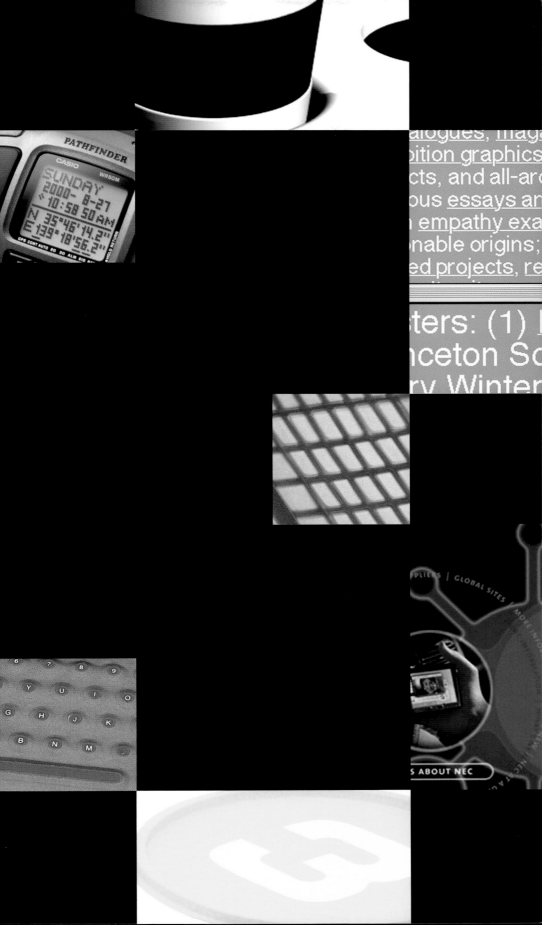

Using Technology

"I don't know why. I just suddenly felt like calling."

We are totally wired–or wireless **(figure 53)**–with cellular phones, two-way pagers, personal digital assistants, laptops, webcams, and electronics that integrate many of these functions into one device **(figure 54)** keeping us constantly in touch with each other. The disdain that some had voiced for the seemingly senseless proliferation and wanton use of such messaging devices sharply diminished in the aftermath of the cataclysmic terrorist attacks on the United States in 2001, when we realized that it was wireless electronics that had kept communications open during the disaster and would continue to provide us with a sense of security and reassurance in moments of uncertainty, independent of conventional grid-based systems. At the same time, the faith that some had placed in the benevolence and efficacy of complex technology was shaken by the ghastly images that revealed an almost-too-easy ability to circumvent technological barriers and commandeer the instruments of twenty-first-century civilization with evil intent. This already had been demonstrated in the less violent but similarly destructive actions of the hackers who had plunged entire electronic networks into havoc with the emission of computer viruses and other infringements on private cyberspace.

Whether we like it or not, and whether we are technophiles or technophobes, our lives are ruled by computers, not so much by everyday communications devices such as pagers and cellphones as by the invisible microprocessors embedded in almost every product we use, from stoves, dishwashers, and microwaves **(figure 55)** to automobiles and cameras. They are there to save us time and labor and supposedly to make our lives much more convenient. We generally don't give much thought to the fact that microprocessors are at work yet we depend on them heavily; we expect our washing machines to turn themselves off when the clothes are dry and air bags to inflate if we hit our brakes suddenly, relying on the sensors inside these products to monitor the slightest changes in activity and respond to them as programmed. All this has become possible because of the increased density and the ever-diminishing size and cost of microchips, which allow them to be used efficiently and economically more and more widely. The technology of embedded micro-processors does not stop externally, since chips now enter our bodies in diagnostic tools that search for the cause of illnesses, and they are implanted for longer periods in prosthetic devices, such as artificial retinas, which are being tested to fill in for damaged membranes. The brilliant technological innovator Ray Kurzweil foresees an even greater role for the human-computer interface in his book *The Age of Spiritual Machines*. He predicts that computers embedded in our bodies will biologically enhance our innate capabilities over the next decades and then propel the evolution of the human organism into its next stages.[27]

For the moment, however, it is our products that are evolving most rapidly, and they are getting much smarter. They are beginning to anticipate our needs. They select the size of paper we should use for our photocopying, let us know how many miles we have left in our gas tanks, and, through global-positioning technology, tell us where we are **(figure 56)**. Machines also are learning how to communicate with each other electronically, and information management is taking over from product hardware as the business of many appliance companies. Web-based communication is used to coordinate the tasks of many different home appliances, regardless of manufacturer, turning them off and on and monitoring them from a central control device, with

54
Teenagers were captivated by the youth-oriented Cybiko handheld wireless all-in-one system with minute keyboard when it reached the market in 2000. A multipurpose device, it includes messaging and chat features, interactive gaming, e-mail (accessed through a computer link), MP3 music (by adding a cartridge), and such other applications as personal organizer, address book, calculator, and English/Spanish translator.

55
The ideographic representations of appliances that appear on Laurene Boym's Glow rug for Handy (2001) recall computer icons and suggest how much these products have been integrated into our lifestyles; we give little thought to the fact that most now rely on advanced computer technology for their basic function as well as for added features that only such technology would permit. The glowing icons mimic the way these products are seen on neon signs, and they have been woven in phosphorescent yarn to replicate the brilliance of neon at night.

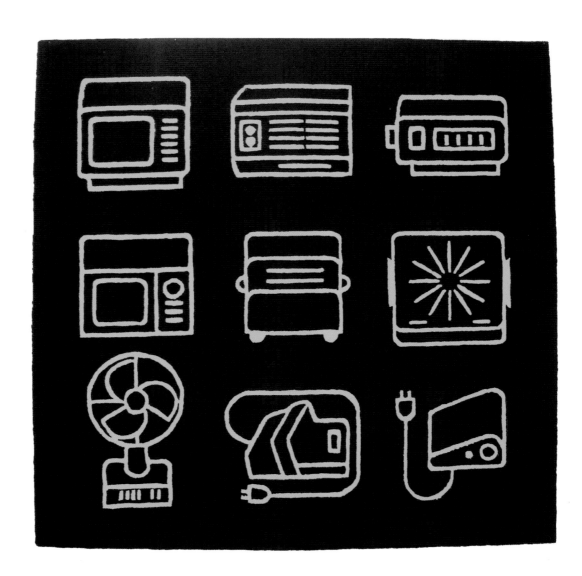

security and energy efficiency along with convenience among the goals of this centralization. Analogous technologies extend to ourselves and our families, with wearable clothing-like devices monitoring our health and web-assisted video cameras monitoring the whereabouts of our aged relatives.

Household products also are becoming mobile and independent as they do their jobs robotically. Friendly Robotic's Robomow can mow our lawns and Dyson's robotic vacuum cleaner can methodically navigate its way around our rooms to clean them efficiently. But robotics have taken their greatest leap for consumers within the toy industry in the form of "independent" interactive creatures. Unlike the cadres of expressionless slave-machines who performed menial tasks for human masters in science-fiction stories, these robots are not faceless but friendly. They are here to serve our emotional needs and not our material ones; instead of being our servants or our slaves they have become our pets and our children. In 1997 the cyberpet Tamagotchi (Japanese for "little egg to hold and love") arrived in the form of a brightly colored flattened plastic egg. Its diminutive LCD (liquid-crystal display) screen monitors the imaginary life cycle of a pet "chick," which demands that we nourish it by pressing the right buttons and rebuffs us when we neglect it, exhibiting good and bad behavior as it grows up. Tamagotchi captivated the marketplace with its personification of human moods and reactions and with its seductive dependency, which forces its owners to maintain a continuing, nurturing relationship in order for its "life" to continue. Tamagotchi technology quickly inspired other devices, notably, the lovable, furry Furby, which speaks its own language, develops as it grows older, and can learn from other Furbys. Sony's plastic doglike creature Aibo ("companion" in Japanese)

and its "cute" and less expensive second-generation offspring, Latte and Macaron **(figure 57)**, also develop their own personality traits. In toys such as these, and the games that bring excitement to Sony's PlayStation and Microsoft's Xbox, ideas that originate in the realm of serious scientific research find early pragmatic and universal application, giving our children and the children within us access to sophisticated technologies that are yet to be applied meaningfully to our own lives.

Long before advanced robotic toys, personal computers, and the Internet saw the light of day, Marshall McLuhan, the controversial cultural thinker of the 1960s, underscored the broader, human implications of electronic technologies with his celebrated epiphany "the medium is the message," published in his book *Understanding Media*.[28] As he predicted, and has frequently been pointed out, the computer, which has elbowed its way into almost every area of our lives since the 1980s, has caused a change in cognition, reversing a mind-set that had been in place for centuries. Digital technology has challenged us to abandon the linear, one-dimensional, left-hemisphere model of our rationalist Western way of experiencing the world for a synchronous, right-hemisphere model that is multidimensional, able to go off in many directions at the same time. McLuhan insisted that instantaneous means of communications would force this change on us since, as he recognized in his later book *The Global Village,* "electronic media themselves are right-hemisphere in their patterns and operation."[29] The World Wide Web, where we start wherever we want and end where it takes us, where we scroll and click and scroll again to lose ourselves in the nether world of cyberspace, validates McLuhan's understanding of modern media.

With all its links and possibilities, the web was the defining technology of the 1990s, regardless of its failure to live up to the economic expectations of many young entrepreneurs. What most distinguishes the web from other mediums in the long history of communications, aside from its complete immateriality, are its links to an utterly unimaginable magnitude of information and its totally democratic "global embrace," in which, as McLuhan envisioned, "the creative process of knowing [would be] collectively and corporately extended to the whole of human society."[30] People in all parts of the world and on all levels of society have logged on, and the web is everyman's tool for getting information, buying goods, and sharing, be it webcam images of the team at an observatory in Antarctica, baby and wedding pictures posted by a commercial photographer, or our own opinions expressed in chat rooms and on personal home pages and web logs.

But however multidimensional surfing the World Wide Web might be, its form of communication is still rooted in text-based thinking. The HyperText Markup Language (HTML) for describing the contents of a web page that the Oxford-educated systems designer Tim Berners-Lee devised when he invented the World Wide Web in 1989[31] is fully embedded in the conventions of print-based communication, supported by such nomenclature as web "pages" and "bookmarks." The earliest web postings that extended his idea of global information-sharing within the scientific community were little more than ordinary typed pages given the linked, interconnective properties known as "hypertext." The designers who brought concern for the standards of graphic design to the web in the mid-1990s continued to support the text model, even as icons, buttons, and other bells and whistles were introduced. Now, dynamic introductory experiences put together by designers with media backgrounds who draw on the increasing sophistication of Flash, Shockwave, and other such technologies go beyond simple messaging and linear presentation of information; we are led, informed, entertained, and seduced by the combination of sound, video, film, animation, and text. But the edgy, open interactivity that created so much excitement during the adolescence of web design has now been overturned by efficient functionality as commercialism and marketing have taken over and redefined the medium. There is a certain sameness about our experiences as we surf the web today, and although we can find outstanding examples of graphic inventiveness designed for commercial clients **(figure 58)** and sites that emphasize simple clarity **(figure 59)**, the conceptual model for many of the sites with the most hits is as economically pragmatic as it is regressive from a design point of view, displaying all the graphic excitement of a supermarket flyer or a tabloid newspaper.

Technology has radically altered the rules of graphic design as it is applied to the web. Print design is predetermined, translated to paper through means of the printing press much as the designer envisions it. Web design is not like that; its appearance is not immutable, since a web page will appear differently on different monitors and with different web browsers, the software programs that allow us to navigate and explore the information super-highway. Moreover, following the intentions of the World Wide Web's creators, variables are determined not by the designer but by the receiver. We can decide for ourselves how we want web sites using HTML to appear on our screens; we can select our own typeface, its size, its color, and its style and

58
Twenty2Product's web site for NEC Global (1998) relied on a hexagonal concept for its lively interactivity and clear functionality. (Produced by Jane Hall; copyright NEC USA, Inc., 1998)

59
The web site of the New York City design firm 2x4 Studio uses simple, linked sans-serif text to tell us who they are and what they do, exploiting cascading features to bring additional, supporting information into view as we want it.

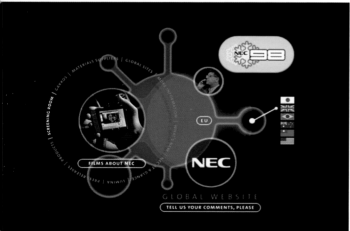

This is the 2x4 web site in which you will find an introduction to the personnel; samples of our projects (including branding and identity systems, advertising and web sites, posters, collateral materials, books brochures and catalogues, magazines, film and video, environmental graphics and exhibition graphics); a list of the artists, musicians, galleries museums, architects, and all-around smart people who are our clients collaborators; various essays and articles; information about the Museu f the Ordinary; an empathy exam; a curious collection of evidentiary material of questionable origins; a messy area dedicated to work in progress, unfinished projects, rejected ideas, unresolved concepts and ther things people we li find us, h oking for a

posters: (1) Lecture series at Princeton School of Architecture; (2) Terry Winters at Lehmann Maupin Gallery; (3) Sam Taylor Wood for Matthew Marks Gallery; and (4)

60
ElectroTextiles'
fabric keyboard inter-
face (2000), made of
metal-coated synthet-
ic fibers, attaches to
a cellular telephone
or personal digital
assistant and relays
electrical surges back
to the device when
the keys are hit. Soft
and washable, the
keyboard can be rolled
up for convenience
and is one of a new
breed of computing
aids and electronic
devices made of soft,
flexible, and even
wearable materials.

choose the color of the background as well. The ability to customize is especially valuable for those with visual impairments, who can adjust the presentation of the content for maximum legibility. We also can choose to exclude the images, beneficial for those with slow modems or those who don't want to spend time waiting for downloads, and this again totally changes the layout that appears on the screen.

In this way, the balance of control has shifted from the individual designer of a message to its receivers. This reversal brought substantial anxiety to professional graphic designers, who, when the reality of the web sunk in, realized that they would have to yield the control over design that they had always considered their birthright. "The first time I saw the same web page through two different browsers at the same time, a feeling of panic swept through me," recalls David Siegel, an early web designer and author of *Creating Killer Web Sites*. "Why should the pages look different? I knew various systems had different color spaces and resolutions, but these programs actually presented the pages in different ways. As a graphic designer, how could I design web pages if I didn't know how they would look? Could I let a piece of software reinterpret my work as it wanted?"[32] For designers, this was a new way of doing business, in which accessibility and usability rather than appearance took precedence, bringing a new level of functional responsibility to their work. Web designers did not take this development lying down; from the start they sought to get around the anarchy of HTML. In order to reclaim appearance as their own domain, they created subterfuges and devised alternative means of structuring the web page so that as many elements as possible could not be easily altered by the receiver.

61
This advertisement for Sony's VAIO computer (2001) portrays the desktop screen as a new artistic medium, suggesting that we throw away our paints and canvases and use digital means to implement the creative process.

Until recently, computer interface has been mostly achieved mechanically through typing, and the introduction of lightweight, fold-up keyboards of plastic and roll-up ones of conductive fabric **(figure 60)** has made this method of data input more readily available for mobile users. With voice-recognition software not yet at the point where it can be applied universally, we are being tantalized with alternative input modes that bring us back to the basics of direct communication by hand and expand what we can do with our technology. Mobile devices using compact pen- or stylus-based data-input systems interpret our own handwriting or use special script for character recognition, as Palm's proprietary software does. Microsoft's Tablet PC design allows us to take notes in ink on the screen just as we would do on a pad, manipulate our handwritten words as if they were text, and then save them, either as an image in the computer's memory or converted to type through handwriting recognition. Desktop systems such as Sony's VAIO Slimtop **(figure 61)** let us draw and paint, and erase if necessary, directly on the flat screen by using a stylus. Pen transmission, another technology that relies on direct interface, also is challenging keyboard input.

Technology does not signify only advanced electronics and the yes/no choices of computer microprocessors; it comprises all the tools that have contributed to the rise of civilization and that continue to civilize our lives today. In applying our skills to the creation of products for practical use, we also are hooking up with older formulas to create new approaches for manufacture, following Tom Dixon's dictum "Rethink." In a move whose popular international success could hardly have been anticipated, the Razor company employed the tools of high-level engineering to reconfigure a

62
Razor updated an
uncomplicated child's
toy into a finely engi-
neered folding scooter
(2000), achieving
unpredicted populari-
ty among commuters
and children alike.

63
Jerszy Seymour
uses a ready-made,
remote-controlled
car mechanism to
put his whimsical
FreeWheelin' Franklin
table for Sputnik/IDEE
(2000) into action
and to direct its
course to anywhere
in a room.

child's scooter for adults **(figure 62)** while reengaging the youth market in a product that had been out of favor for decades. The scooter runs smoothly and impeccably on colorful in-line-skate wheels and can be folded up to take into the office. Razor's scooter and those that imitated it were positioned as an alternative means of transportation for getting to work and around the city and also had health benefits for city dwellers who were encouraged to use it for exercise, at least before small electric motors were added to speed them up in traffic. A more goofy adoption of available technology is Jerszy Seymour's FreeWheelin' Franklin table; an amalgam of a plastic top emblazoned with racing-car numbers mounted on wheels and guided by the motor of an off-the-shelf, remote-controlled toy car, the table can be directed to transport its contents to any part of a room **(figure 63)**.

For his part, the English inventor Trevor Baylis turned to the technology of early hand-cranked clockwork record players to devise a radio that would run for an hour on less than a minute's hand turning. Baylis conceived it in response to accounts of the lack of success of AIDS campaigns in South Africa because information could not be brought to remote populations due to the high cost of batteries and absence of electric lines to power radios. His invention, now marketed in several colors as the see-through plastic Freeplay radio **(figure 64)**, proved of immense value in third world and war-ravaged countries. Like so many devices created for particular situations, it also has proven valuable for the entire population for use in emergencies during power outages and regularly on boats and in vacation areas that are not connected to the electric grid. The same simple technology was subsequently adapted by Sony in its Audio Beacon radio. Freeplay has used such simple

64
Revisiting the basic
technology of spring-
driven clockwork,
Trevor Baylis applied
it as a power source
for his radios with
windup handles,
which first appeared
in 1996 and are now
distributed in various
sizes, formats, and
colors by Freeplay.

65
The FreeCharge, a
windup charger devel-
oped by Freeplay in
conjunction with
Motorola (2002), is
ironically intended to
supply power to
cellular phones and
other advanced elec-
tronic products.

66
Arik Levy admits
that his "virtually
real" Infinity lamp
(1999), sold as a
videocassette, is a
contradiction. "It's a
light but it's not an
object," he explains,
it's "technology."

technology for other common devices such as flashlights and, ironically, for pocket chargers for high-tech products, such as laptops and cellphones (figure 65). Parallel to this, photovoltaic technology has harnessed the energy of the sun to power many different products, from our common pocket calculators to entire energy systems for communication and lighting in countries where grid-based electricity is not generally available.

Designers also have returned to the comfort of earlier technologies through metaphor, especially in the field of lighting, where many new devices have brought the bulb out from under the shade and into view. Arik Levy's Infinity lamp for Snowcrash (figure 66) is a virtual light bulb sold as a videocassette, which shines from a digital screen or through video projection. His concept has the potential to achieve truly democratic distribution, because as Levy explains, "if it was broadcast millions of people could have my lamp in their homes."[33] Ingo Maurer too has reconsidered the "bare" bulb in many of his works, and in his Holonzki lamp he used a hologram to remind us of the underlying constant of our electric lighting (figure 67). Marcel Wanders has gone back further, with his B.L.O. candle lamp, to remind us of earlier forms of illumination (see figure 28).

But technology extends beyond the domain of hardware to the realm of materials and processes, which are now the focus of increasing interest among designers. Recent discoveries have shown how substances can be made to change their personalities, so that, as Ezio Manzini pointed out in *The Material of Invention,* a book that has inspired a great deal of design discussion over the last decade, materials are "increasingly difficult to

67
Ingo Maurer chose the advanced technology of hologram photography to project a bulb when his Holonzki lamp (2000) is turned on, conveying a metaphorical reminder of the simple origins of electric lighting.

68
Many new materials can change their physical properties in response to changing circumstances. Josh Owen harnessed the color-changing ability of thermochromatic liquid crystals for his heat-sensitive tray and glasses for DMD (2002). Physical contact, the proximity of hot or cold objects, and ambient temperature all change the appearance of these surfaces.

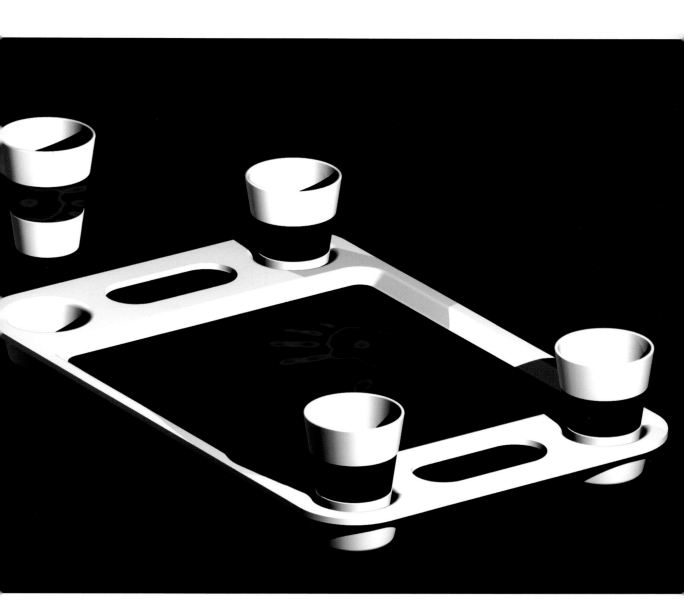

define in simple categories that we can say have been acquired once and for all. The only way to describe the material is to consider it as a system capable of performance; thus we [must] speak of a 'material,' not by defining 'what it is,' but describing 'what it does.'"[34] Many materials now have integrated responsive technologies, which allow them to change their shape, color **(figure 68)**, hardness, and other physical characteristics according to circumstances, as well as "memory," which allows them to return to previous configurations after being disturbed. In one large-scale demonstration, readers of *Wired* magazine in January 2001 were surprised to see that their fingers left temporary yellow marks on its green cover, tagged "Touch me all over," which was printed with a thermochromatic ink sensitive to body heat.

The expectations that materials should look, feel, and behave in very specific ways, one of the cherished principles of modernism, have been overturned, especially with the many different composites now available. We naturally think about materials in clichés—fibers are soft, ceramics are fragile—and expect that their application be limited to narrowly defined areas. But Marcel Wanders's Knotted fiber chair **(figure 69)** is hard, and Kyocera's high-fired zirconia ceramic knives are extra-sharp and durable. Fibers do not customarily create a rigid material, and unless slung like a hammock, a knotted or woven fabric could not normally support a person. Yet by impregnating it with epoxy, a fabric created of an up-to-date, high-tech fiber with the age-old hand-knotting macramé technique can be made hard and rigid, strong enough to serve as a chair. In this way Wanders arrived at a product that is not only totally new but that also totally confounds clichéd expectations of its material.

69

The Knotted fiber chair that Marcel Wanders designed for Droog Design (1996) combines macramé, an age-old hand process, with new materials technology. Using the synthetic fiber Aramid around a carbon core, he forms a soft knotted fabric and fashions it into the shape of the chair, creating the seat first and then the legs.

After being impregnated with a nontoxic epoxy resin and stretched in a frame, the once-limp form hardens, relying on gravity as an assistant in the process, and becomes strong enough to support a person's weight.

70

Werner Aisslinger's Soft Cell chairs and lounges for Zanotta (1999) introduced soft plastic gel as a material for consumer home products.

71
Hella Jongerius's
soft polyurethane
Pushed Washbasin for
Droog Design (1997)
introduced a new sub-
stance and a new feel
to sanitary fixtures.
Replacing the conven-
tional cold, hard
porcelain with soft
plastic, she allowed
the final shape to
evolve from the vari-
ances in thickness of
the material and cre-
ated the rolled rim by
pushing the form
inward.

Plastics are perhaps the most variable of familiar materials used in design; they can look, feel, and be manufactured in many different ways, and they "do" many different things. Gels, for example, are soft polyurethane substances that look like they are filled with fluid. They yield when we sit on them as pressure is equalized across the surface and return to their original form when we get up, but they do not leak or break; commonly used as padding to cushion the body in medical supports and in bicycle seats, gels recently have been introduced into furnishings. They serve as both a utilitarian and an emotive decorative substance in Werner Aisslinger's Soft Cell transparent lounge chair (figure 70). Soft polyurethane, which is used industrially for making molds because it is easy to bend and remove the casts, also has captured the imagination of Hella Jongerius, who uses it perversely to transform products that are conventionally hard and rigid, such as vases and sinks, into objects that are soft and flexible (figure 71).

Another approach that has been engaging designers more actively is biomimicry, in which the workings of nature are imitated in devising new materials, processes, and products. Nike, for example, has studied the hooves of animals as a natural model for the configuration of the soles of running shoes while many different natural reactive substances are being investigated as models for the creation of intelligent materials. But biomimicry's broadest application may be in the field of material ecology, where natural processes can lead the way to revising our understanding of substances so that by following biological models, we can reduce our harmful impact on the planet.

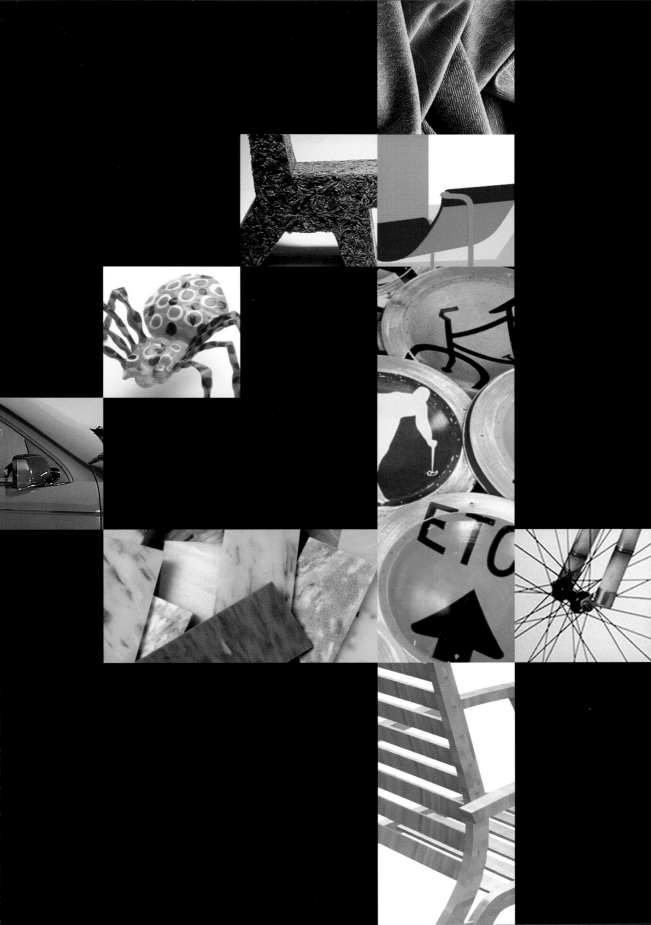

Being Responsible

Many more designs than we would expect are not fully tested in the field under the physical and social conditions in which they are to be used. This was the problem with the Florida butterfly ballot that was contested in the presidential election of 2000. This TV demonstration shows the confusion of arrows and listings that caused some voters to cast their ballots for the wrong candidate. (©AFP/CORBIS)

No one who reads newspapers or tunes into the media could be unaware of how design decisions can affect not only our individual lives but also the fabric of our nation. The debacle of the voting controversy in Florida during the presidential election of 2000, for example, when the form of the paper ballot left many voters unsure of which number to punch for their candidate **(figure 72)** and when flaws in the VotaMatic machine resulted in the incompletely punched paper chads,[35] pointed out what can happen if poor and untested design is presented to the public. At the same time that Americans were becoming aware of the political implications of poor design, newspapers were filled with articles about how safety could be affected by design, witnessing the recall of millions of Firestone tires that were said to have contributed to rollover deaths in the Ford Explorer. In its reporting, the *New York Times* was able to track down a series of "compromises" accepted during the development of this popular sport-utility vehicle that had put cost savings and the need to bring the Explorer to market quickly ahead of its stability and safety.[36] These, of course, are not the only instances in which problems in everyday objects and everyday experiences have been pinned on the acceptance of poor design—the notorious VCR, which ceaselessly flashes 12:00 because we can't figure out how to set it, is a common and frustrating example.

While we have been forced to become responsible in certain areas, for example, through product-safety legislation and with the introduction of seat belts, air bags, and other such features on automobiles, our health and our natural resources as well as our institutions and our overall safety are repeatedly at the mercy of a system that does not make responsibility in design a national or international priority. Although the positive economic and social repercussions that can come from a national effort toward the improvement of basic design have been recognized, the United States has not embraced this on a government level as have other countries (especially Great Britain, which includes design studies as part of its national school curriculum) but relies on corporations, museums, and professional schools and societies to further efforts in this direction.

Although it would seem obvious that throughout each stage of design products would be studied both in the designer's studio and in real-world situations so that problems of functionality, usability, and safety could be worked out, this is often not the case, as the psychologist Donald A. Norman has shown in his eye-opening book *The Design of Everyday Things*. "What about my inability to use the simple things of everyday life?," he complains. "I can use complicated things. I am quite expert at computers, and electronics, and complex laboratory equipment. Why do I have trouble with doors, light switches, and water faucets? How come I can work a multimillion-dollar computer installation, but not my home refrigerator? While we all blame ourselves, the real culprit—faulty design—goes undetected. And millions of people feel themselves to be mechanically inept."[37] Norman lays some of the problems of the lack of usability of everyday objects on the arrogance of designers, who assume we are able or willing to follow detailed instructions about how to use them; and on manufacturers, who instead of simplifying products add more and more features to individualize them in the hopes of making them more marketable. He also shows that many common problems with products could be avoided if designers followed very simple principles, such as providing visual clues as to how an object works, putting controls into the natural sequence in which they are to be used, and then giving feedback so that we know if our actions have been successful. Faced with poor design, we often are presented with new products created to remedy the problems of the earlier ones, such as lumbar supports for uncomfortable chairs or ergonomic accessories to ease the strains of working in unnatural positions that result in carpal tunnel syndrome.

If we extend our vision of responsibility beyond individual objects to their societal context, the primary concern in design today must be universal access, to bring the basics of industrial design—sanitation, electricity and communications, adequate tools—to developing nations. This is a moral and not a financial issue, and even in periods of international economic downturn, the disparity between haves and have nots leaves us no choice but to work toward alleviating the desperate conditions under which so much of the world's population lives. This is a goal that could be achieved with a relatively small outlay if the roadblocks of bureaucracy and politics could be overturned. The introduction of appropriate design and technology in concert with the needs and wishes of local populations can open up the process of providing both locally desirable and culturally specific utilitarian products to those who require them and access to the means to make products for themselves and for sale.

Craft economies in developing nations have been battered by the evolution of technologies and changes in consumer preferences (from pottery to plastics, for example). The choice for local craftspeople who remain outside of the

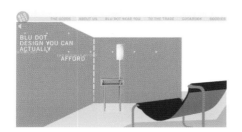

BLU DOT
DESIGN YOU CAN
ACTUALLY
AFFORD

73
IKEA manufactures
its products in large
volume, which, along
with knockdown
packaging, results in
affordable prices as
demonstrated by its
painted Edvard chair
designed by Nike
Karlsson (1998).

74
Blu Dot's web site
clearly states its
philosophy of selling
"design you can
actually afford."

global manufacturing system seems to fall into two categories, each of which impacts their economy in a different way. On the one hand, designers can devise new products for local manufacture that will meet the requirements of indigenous populations, and on the other, the government and other organizations can put industries in touch with tourist or ethnic-export markets, where traditional shapes, motifs, and decorative designs can be used in new ways to bring income to the locale. Local, national, and international initiatives—both commercial **(figure 86)** and nonprofit (such as that of Ten Thousand Villages, an organization that commissions products from village craftspeople for sale abroad)—have been credited with creating new markets, but the task of turning craftsmanship into new products and international sales calls for the mediation of cultural differences between makers and their distant customers. As much as industrial societies sentimentalize the hand craftsmanship that has been lost to them but is still in evidence elsewhere, it is unrealistic to think that large pockets of traditional craft activity will survive much longer. As the Indian designer and critic Martand Singh has pointedly shown, "We can look at an earthen pot and say, isn't this a great object, and we must sustain more and more of these great objects, because that is ethnic chic. And ethnic chic does arise out of our inability to create new horizons for craft. And our inability to accept the fact that the man who makes [the pot] wishes to become like you and wear a watch which is industrially produced and shoes which are industrially produced and live in a house with a fridge, have a fan or air conditioner or whatever, and none of this is related to traditional craft."[38] We must recognize that these same craft abilities have wider application to the field of skilled industrial production.

Even in industrialized countries, quality, well-designed products have not been consistently available to those who are less well off and who must rely on the cheap and often poorly thought-out merchandise sold by mass or cut-rate marketers. Although they do not aim specifically at the lowest population segments, some major retailers, such as Habitat in Great Britain and IKEA, offer well-designed and reasonably priced objects in an effort that recalls the democratic ideals, and sometimes the forms, of Good Design **(figure 73)**. Blu Dot, an American firm, also creates flexible, knockdown furniture and accessories made for easy, economical distribution in a consistently more progressive style **(figure 74)**.

Sustainability—the creation of environmentally sound products made from renewable resources—is the overarching concern for those who would be both forward-looking and responsible in the creation of design today. The emergence of a critical mass of activism in the United States for all aspects of the environment dates back to Earth Day 1970, when twenty million Americans marched throughout the nation to express their support for efforts to restore the well-being of the planet. Efforts to coordinate international environmental activity culminated at the 1972 Stockholm Conference on the Human Environment, which ceded control of that initiative to the United Nations. At first the proposed ecological solutions leaned heavily toward the cleanup of hazardous conditions and recycling; at the same time a retrogressive, anti-industrial stance emerged, manifested by a back-to-nature commune movement and such do-it-yourself publications as the *Whole Earth Catalog* (from 1968). Today the viewpoint is more proactive, focusing on the need to cut pollution, reduce the consumption of fossil fuels, and make

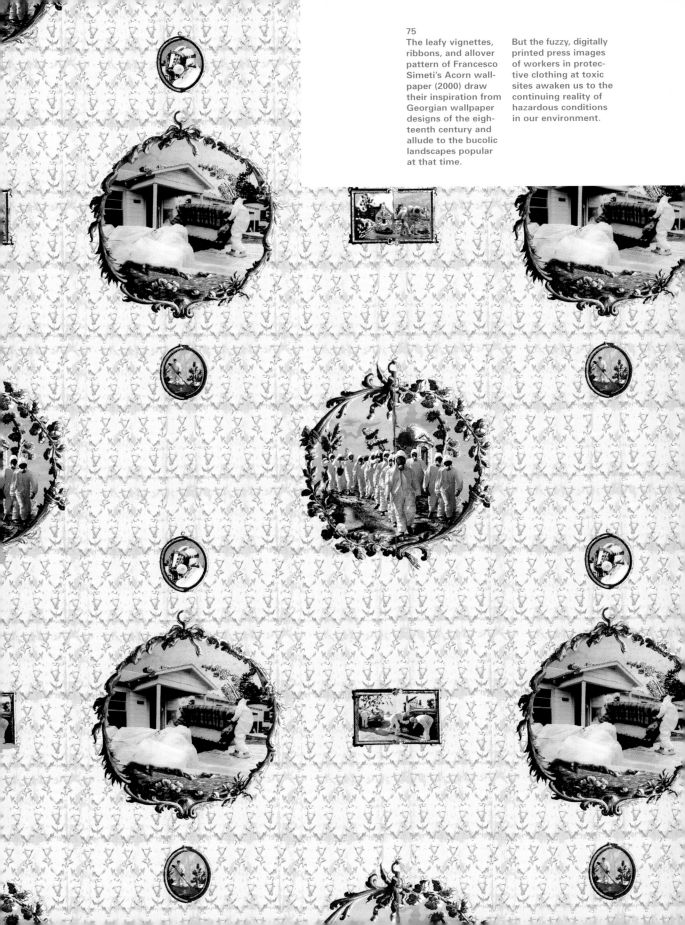

The leafy vignettes, ribbons, and allover pattern of Francesco Simeti's Acorn wallpaper (2000) draw their inspiration from Georgian wallpaper designs of the eighteenth century and allude to the bucolic landscapes popular at that time.

But the fuzzy, digitally printed press images of workers in protective clothing at toxic sites awaken us to the continuing reality of hazardous conditions in our environment.

When it sees red, it charges.

76

Toyota's Prius was the first hybrid family car to be sold in the United States (2001), responding to federal requirements for automobiles with lower emissions levels.

As this advertisement shows, the Prius runs on dual energy sources, a gasoline engine and an electric motor, which alternate according to the particular driving circumstances. For normal driving, the gasoline engine is in use, but the electric motor takes over when the car slows down or idles, and the energy produced by deceleration is captured and converted to charge the battery. This results in fuel savings and lower pollution.

industry sustainable while recognizing that recycling efforts do not solve the underlying issues of the destruction of our resources. We must consider the impact of products in all stages of their life cycle, from materials and manufacturing to packaging and disposal, in order to assess their environmental impact. But we still have an uphill environmental battle before us, as was demonstrated by the uneasiness of the United States at the prospect of signing the Kyoto accord on global warming in 2001—and as Francesco Simeti reminds us ironically in his wallpaper showing muted scenes of toxic contamination removal amid decorative vignettes in an eighteenth-century format (figure 75).

In the 1970s, ecological design was often self-conscious, calling attention to itself—and to what was then its fringe, often deprecated as "hippie," advocates. After thirty years, ecological design has been mainstreamed, and responsible design generally eschews being obviously so as it reaches into all areas of the economy. Toyota's Prius, the first hybrid five-passenger car to be sold in the United States, combines an electric motor and gasoline engine to reduce energy use and harmful emissions but looks and performs very much like any other car on the road (figure 76). Yemm & Hart, an American company that recycles tires and plastics for construction materials, uses 100 percent postconsumer polyethylene from bottles for their Origins line (figure 77). They joyfully exploit it by artfully blending the flakes of shredded plastic and molding them into panels with multicolored, random patterns: "By making a recycled material look and feel desirable," and by adding several colors to plastics, which usually have a single color, "the material takes a step forward, changing the use of recycled materials from a cause to a common place."[39] But

77
Yemm & Hart recycles plastic bottles for its Origins surfacing products, which have colorful, decorative patterns that extend throughout the thickness of the material. In order to produce these materials, the bottles must be collected, sorted by color and shredded, and then blended into specific color mixtures.

some designers still want to broadcast their commitment to environmentally friendly design in order to capture a concerned audience or to encourage others to understand its benefits. The design firm Bär + Knell works with the German recycling institute to promote the use of recycled materials in manufacturing. Their mass-produced but one-of-a-kind furniture proudly announces its source in recycled plastic containers by incorporating the products' labels into their surface **(figure 78)**. At the same time, they have introduced recycled bottles anonymously in a multicolored confetti pattern or in solid colors by rethinking the seats of a line of office chairs designed by Egon Eiermann in the 1950s and manufactured by the German company Wilde + Spieth.

Investigation into new materials and processes has uncovered sustainable sources that can replace conventional ones that have proved environmentally harmful. Biopolymers, such as those made from potato starch and cornstarch, have the same characteristics as certain plastics and can be molded and formed in the same ways, but they are fully biodegradable. Cargill Dow's NatureWorks plastics derived from corn and other plants are made into a variety of fibers, packaging, and other forms but will break down naturally when they are discarded, as will KidTech Tools' Magic Nuudles colorful building material created from cornstarch. Plants in their unprocessed forms are readily available sources for sustainable production. Bamboo, one of the planet's most vital plants, is virtually inexhaustible while gourds, bark, and wild (or other easily cultivated) plants are readily available for fiber use. Bamboo and rattan furniture have long been at the center of craft industry in Southeast Asia, and new uses for these plants are being introduced. Bamboo

78
Bär+ Knell, a group
that works with
the German national
recycling agency to
create environmental
awareness, molds its
furniture from used
plastic packaging,
flaunting the origins
of the material with
the distended labels of
the products that
shroud their surfaces.
The firm's objective
is to remind us that
"packaging is not just
waste; it is a valuable
raw material."

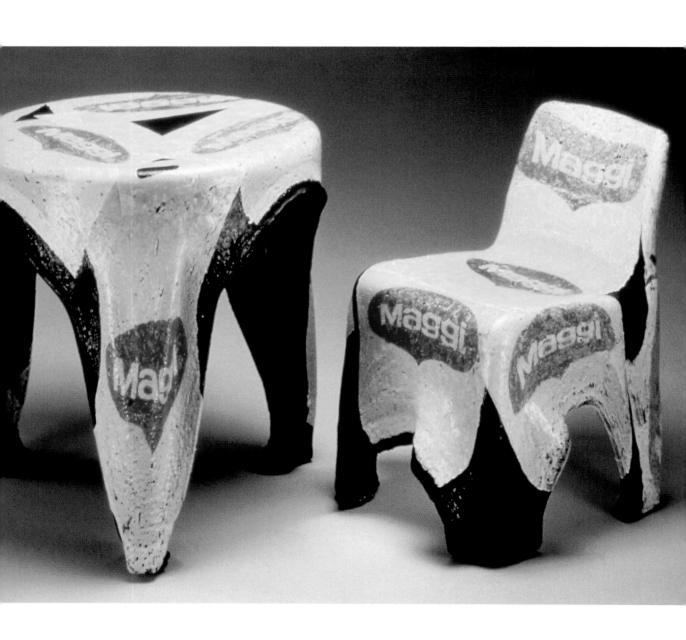

composites have wide possibilities for flooring and are being introduced as well for surfboards and skateboards. Bamboo is again being considered for bicycles both decoratively and structurally (figure 79), as it had been in the late nineteenth century, replacing their metal frames on account of its great strength (greater than steel for tension). The architect Michael McDonough, cofounder of the Bamboo Research Initiative at the Rhode Island School of Design in Providence, brought high-tech considerations to design in bamboo with his laminated-bamboo chairs (figure 80). Wood, when farmed properly, is another renewable source for ecologically sound manufacture. The Finnish designer Ritva Puotila employs ecologically produced and organically dyed spun-paper yarns for her sustainable Woodnotes carpet collection.

The greatest focus has gone, however, to the utopian thinker William McDonough (no relation to Michael McDonough), who has become the poster child for responsible, environmentally sensitive, and sustainable design. An American architect of outstanding eco-friendly buildings for such firms as Nike and the Gap, McDonough has been celebrated both professionally and popularly; he was named 1999 Designer of the Year by *Interiors* magazine and was featured as one of the Heroes for the Planet in *Time* magazine the same year.[40] In 1995, in association with his partner Michael Braungart, a German chemist, he joined a group of well-known architects and designers in launching a new line of upholstery fabrics for DesignTex. But instead of accepting the company's seemingly ecologically sound suggestions to use natural cotton and fibers made from recycled plastic bottles for his designs (he questioned cotton because of the large amounts of herbicides used in its production and the recycled plastic because of the harm its unknown

79
Flavio Deslandes built his first bamboo bicycle prototypes when he was an industrial-design student in his native Brazil, devising his own manufacturing technology and using the country's plentiful natural resources as his raw material.

Since moving to Copenhagen, he has begun to manufacture city bikes in sleek, improved designs, which combine elegant, well-engineered, steel joints with the plentiful, strong, and sustainable material bamboo.

80
Exploiting bamboo's tensile strength, the architect Michael McDonough created ultralightweight chairs that reflect aspects of both Western and Eastern design traditions.

components might bring to the human body), he insisted that the entire manufacturing process be reevaluated to make the production of his textiles environmentally safe and sustainable. This extended beyond the selection of a combination of organic animal (wool) and plant (ramie) fibers, which together would create a fabric that was cool in summer and warm in winter, to establishing strict ecological criteria for the choice of the dyes and other chemicals used in the manufacture, the monitoring of energy and water consumption and quality during the production process, and even consideration of the lubricants used on the machinery. The resulting fabrics (figures 81, 82) were executed according to what they have named—and trademarked—the McDonough Braungart Sustainable Design Protocol. As product conceptualists, they categorize products based on environmental values; after eliminating those that are harmful (and thus should not be made at all), they divide those they consider viable into two different spheres of sustainability in what they call "cradle-to-cradle" life cycles. Organic "Products of Consumption," everything from clothing to packaging, are meant to be biodegradable; after use they can be thrown on the compost heap and returned as nutrients to the soil. Technological "Products of Service," such as automobiles and television sets, follow their own closed cycle of use and reuse; when disposed of, many of their elements can be remanufactured into similar products rather than "downcycling" them into a sphere of lesser commercial value.[41]

Design for disassembly and remanufacture is a major initiative for sustainability that has been adopted as a national policy in some countries. In Germany, for example, products such as home appliances and electronics

81
William McDonough's
textile collections
for DesignTex follow
the strict manufactur-
ing procedures of
his "cradle-to-cradle"
life-cycle protocol of
product sustainability
but appear to the user
like any other fabric.

82
Moss, from the third
collection by William
McDonough for
DesignTex (2001),
comes in fourteen
colors, is fully biode-
gradable, and is made
without harmful
dyes or production
processes.

must be designed in a manner that allows for easy disassembly when they
are no longer needed; their individual components are either readied for reuse
and remanufacture or their materials are easily separated for recycling, having
been clearly identified during production. Packaging, too, must be minimal,
devoid of advertising, and taken back by the manufacturer. This approach has
been extended on an even larger scale to automobiles within the European
Union with its end-of-life vehicle directive. By 2007 all European
automobiles must be made so they can be returned to the manufacturer for
convenient disposal processing; they will be taken apart when their useful
lives are over and broken down into components that can be remanufactured,
recycled, or disposed of properly **(figure 83)**. Remanufacturing also is being taken
up globally, as much to keep costs down as to be green and reduce landfills.
Xerox designs its copiers so that the parts can be reused or remanufactured
while Kodak collects "disposable" cameras and reuses elements from them
in new ones. American automobile manufacturers also are investing in
disassembly plants to recapture usable elements from earlier models.

Manufacturers often use leftovers and production waste, such as trimmings,
to make additional products, as Rossignol does with snowboards and as
many paper manufacturers do. Shredded plastic bottles, a major consumer
disposable, have found a multiplicity of uses, including the making of
insulating clothing fibers such as Polartec fleece. A simpler approach to
recycling extends the useful life of materials and products with little physical
change: Boris Bally reclaims large metal traffic signs and makes them into
furniture and vessels in his "Urban Enamels" series **(figure 84)** and the Swiss
company Freitag collects tarpaulins from European trucks and recycles them

83

Design for disassembly, or more descriptively, recycling and remanufacture, has become a major environmental initiative in many European countries. With regard to automobiles, it is now a directive of the European Union and specific goals for the amount of materials in an automobile that must be reused have been set for the next decades. In order to succeed in this, life-cycle planning must occur: The design of each automobile must be conceived from the outset with an understanding of what will happen to all of its components when its useful life is over, and all materials must be clearly identified so they can be separated for reuse or recycling.

This photograph of a BMW X5 shows the plastic components that could be recycled from it in 2000. The process of expert disassembly begins when the last owner returns a car to a dealer for free-of-charge recycling. Its fluids are removed and the engine is reconditioned for high-quality reuse.

Glass is separated from plastic and the plastic further separated into its various types for recovery. The metals are shredded and then separated into their ferrous and nonferrous components for further processing. This concept is being extended to the design of many other consumer products.

84
Boris Bally collects the raw materials of his "Urban Enamels" series–used traffic signs–and transforms them into bowls and other products that very clearly display their origins for decorative and communicative effect.

85
The simplest approach to recycling extends the useful life of resources and products with little or no physical change to the material at all. Freitag turns tarpaulins from European trucks into their Freeway shoulder bags, creating unique products that satisfy today's passion for individuality and customization.

86
Using recycled tin cans as their raw material, craftspeople in Zimbabwe create these diminutive, highly decorative animals, which are then painted by local artists. The American firm Mbare markets them internationally and returns a percentage of the proceeds from their sale to an artists' trust.

into their Freeway shoulder bags **(figure 85)** and accessories, which have earned a cult following.[42] Craftspeople as far removed as Zimbabwe and Mexico collect tin cans to use as a raw material for making toys and basic products for the home **(figure 86)**. Jurgen Bey in his Gardening bench for Droog Design suggests a way that we might make even simpler products with virtually no intervention in the life-cycle process. He has pressed hay and leaves with resin to form nondurable furniture, which is naturally recycled when exposure to the elements makes it no longer suitable for use **(figure 87)**. Recycling and reuse are still the most practical areas for public intervention in green design because anything that can be recycled avoids, if only for the moment, adding to civilization's landfills.

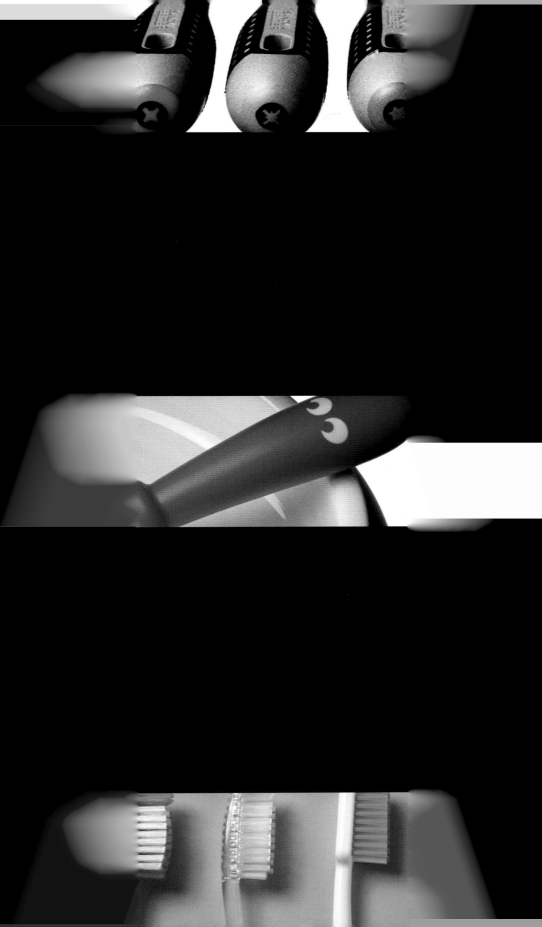

Serving Individuals

88
Backpacks and suit-
cases with handles
and wheels are
examples of products
that aim at the inclu-
sivity of universal
design, making the
transporting of heavy
loads easier for all.

The Class Act
schoolbook backpack
from the retailing
cooperative REI (2001)
comes with in-line
skate wheels and a
T-bar handle for easy
rolling and with
strong harnesses and
straps for carrying
on the back.

Nothing has had as much impact on the human side of design in the United States as the passage of the Americans with Disabilities Act of 1990 (ADA). More than just mandating barrier-free accessibility to public accommodations of every sort, the act established the principle of universal inclusivity. It legislated equal access to all facilities available to the public instead of creating a separate category of accommodation for those with physical and mental disabilities. The mandate to mainstream those who were blind or wheelchair users, for example, forced architects, designers, and social planners to reconsider the configuration of architectural spaces, particularly entries, corridors, and toilets; transportation facilities; and communications media. While the most immediate and most visible effect of the act was the creation of ramps on buildings throughout the country, the impact was all-encompassing.

Unwittingly, the ADA brought enormous benefits to the public at large. The simple changes that accrued and made everyone's life easier, such as larger, clearer signage that was much easier to read and the convenience of ramps or curb cuts for those with baby carriages or luggage on wheels—itself a product with universal applicability **(figure 88)**—pointed out the values that ADA reassessment of design could have for the entire population and pushed to the forefront the universal design, or design-for-all, movement. Following the spirit of the ADA, universal design is an approach that is inclusive, that aims to create facilities and products that everyone can and wants to use rather than those that are directed to particular segments of the population and that set them apart from the rest of society. It recognizes that all of us who today may be able eventually will need some of the accommodating features now found

in design for special needs, and thus making products universally accessible is both a logical and a sound social and economic goal. The range of universal design can extend even further when it brings products closer to the abilities of those with greater needs even if it does not reach them, since it then allows products to be adapted more economically and efficiently if required. But universal, and transgenerational, design as we know it today has limits; it is directed to the generic adult population, without gender or cultural distinctions, and excludes children, often purposely so for reasons of safety or because they may not be developmentally able to use certain types of products. Design for children is a separate area **(figure 89; see figure 21)** with its own universal design and accessibility standards.

Giving thought to including as much as possible of the technology and the elements that individuals who are left handed, aged, or impaired require at the beginning of the design process is the secret of universal design; choosing levers for door handles **(figure 90)** instead of the more difficult to maneuver round handles commonly used in American houses is a good example of a universal application about which most people would not think twice. Ameriphone's Dialogue ER telephone **(figure 91)** includes features that might well be desired by all users: big buttons for manual dialing, photo-display for speed dialing, and variable sound amplification. But these same features serve to enhance the abilities of those with special needs, impaired eyesight and memory, and hearing loss. In addition, the telephone has an emergency-response system of prerecorded dialing, which brings reassurance to seniors and those who live alone that they can press a button and automatically alert family or the authorities should an emergency occur.

90
A simple universal-
design solution is the
use of levers instead
of knobs for door han-
dles. Marc Newson's
aluminum DH handle
for Erreti (2001) is
even more universally
adaptable; it snaps
easily from a left-
handed to a right-
handed to a vertical
position.

89
BabyBjörn's plastic
dinner set (2000) was
designed specifically
for the abilities of
young children. The
clover-shaped plate,
with its raised edges,
aids children in
scooping up food
while the spoon, with
a short handle that
works equally well
with the left or right
hand, accommodates
their developing
motor skills.

91

Although Ameriphone's Dialogue ER (for emergency response) telephone (2001) was conceived to help people with special needs, its functions are universally applicable and can serve a wide population. Taking a cue from the Big Button telephone designed in 1990 for AT&T by Henry Dreyfuss Associates, the Dialogue ER has large buttons that help those with failing eyesight but is stylish enough to be included in any home. In addition, it has six photo displays for programmable speed dialing, of use both to those with failing memory and to children, who can easily make contact simply by recognizing the picture of the recipient. Its most significant feature is a preprogrammed automatic dialing and messaging system, which alerts relatives and other contacts sequentially in case of an emergency and then activates a two-way speaker phone to find out what the emergency is, a feature that would be of value to anyone who lives alone, regardless of age or ability.

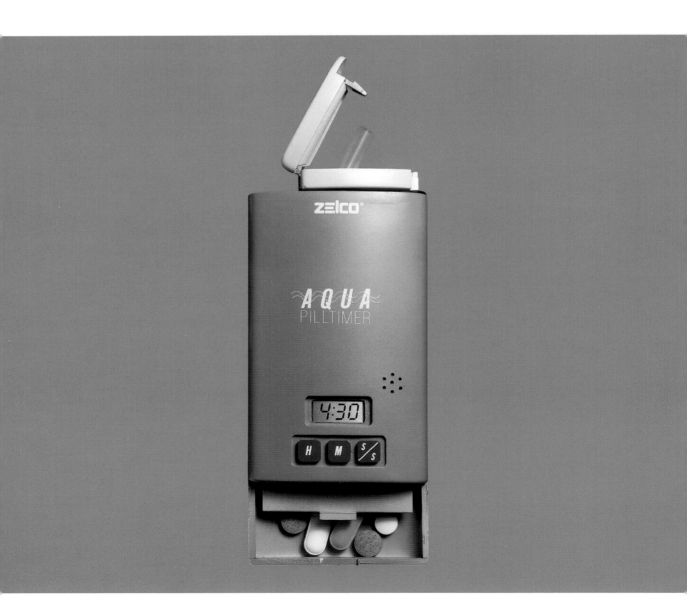

92
Zelco's electronic
Aqua Pilltimer lets us
know with an alarm
or a beep when it is
time to take our
medicine and makes it
convenient by carry-
ing a water supply
and straw along with
our pills. Its features
are equally applicable
to those who need
special help in remem-
bering to take their
medications and
those whose busy
lives distract them
from their schedules.

Likewise, Zelco's Aqua Pilltimer **(figure 92)**, which holds water and pills
and has a programmable timer that reminds us when it is time for our
medicine, is a convenience for anyone who takes medicines regularly but also
works to the benefit of those who may be forgetful in old age. OXO's Good
Grips kitchen utensils and the large range of related products **(figure 93)** that
followed upon the success of its first line designed by Smart Design and
introduced in 1990 speak for the viability of universal design, and even the
branding value of a design element added to accommodate disabilities. These
utensils have a signature thick, black, nonslip handle with a flexible indent at
top, which makes gripping easier for those with arthritis and other grasp
impairments (for whose capabilities they were conceived), but work equally
well for everyone. Many purchasers do not even consider the benefit of the
grip when they buy these products but choose them for their weight, feel,
and materials, their quality of manufacture, and their style without knowing
anything about their origin in design for a special need.

Ergonomics and human factors, two allied fields that are integral to design
today, work together to make products that serve specific needs for
individuals of all capabilities. The science of ergonomics focuses on creating
products in a way that will make them more effective, efficient, and easier to
use and more comfortable for the user **(figures 94, 95)**. One of the tools of
ergonomics is human-factors engineering, which takes into account the
physical and psychological needs of the user, relying on anthropometric data,
the measurement of the human body and the way it functions, to improve the
suitability of a design. Designers frequently consult anthropometric data
published in journals and in the series of charts in the *Humanscale* statistical

93
OXO brought a concern for functionality and universal design to simple tools that had not been approached in this way before. After the success of its Good Grips line of utensils, the firm launched other products including its Salad Spinner (2000), which is equally useful for those who are right- or left-handed and takes little effort to press and turn the spinner. This product was conceived by business students at Babson College and designed by Human Factors Industrial Design.

94
Designed by Sweden's Ergonomi Design Gruppen, long standing leaders in the creation of ergonomic products and those that accommodate physical impairments, their Ergo screwdrivers (2001) belong to a continuing series the firm has conceived for Bahco. The handles are carefully engineered to work with reduced muscle stress, which greatly increases work efficiency and comfort.

95
Each handle of Tupperware's E ("ergonomic") Series of kitchen knives (1998) is unique, designed by Ergonomi Design Gruppen in accordance with human-factors research that suggested the optimum shape required for each of the tasks to be performed.

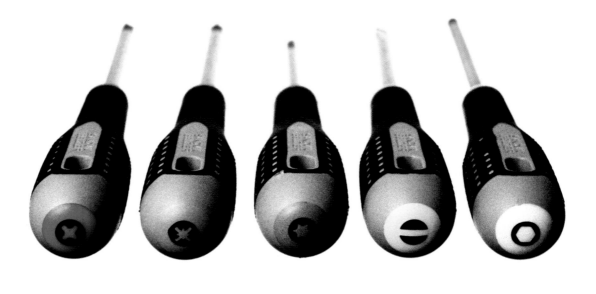

96
There is no way
to tell if these
toothbrushes are
created with true
ergonomic research
or use "ergonomic"
elements simply
for style.

97
Oral-B Laboratories'
CrossAction tooth-
brush, designed by
Lunar Design (1999),
was the outcome of
extensive documented
research into the
ergonomics of brush-
ing and the bacteria-
fighting action of head
design so that the
benefits of dental care
could be maximized.

surveys[43] as a source for information on standard human measurements, which aids them in deciding on the shape and relationship of the elements of their products before they are tested, although these are general data based on averages. Human-factors engineering has most steadily been applied to design for the workplace to ease the strain of performing repeated tasks endlessly on the assembly line, sitting in one position for a long amount of time, or working at a keyboard. Office furniture especially has been a recent focus of this research. Herman Miller's Aeron chair designed by Dan Chadwick and Bill Stumpf in 1994 set the industry standard for ergonomically conceived office chairs, and many other manufacturers have followed their lead, bringing out second- and third-generation concepts of comfortable office seating.

The aura of ergonomics also has invaded many areas of design for personal, everyday use. There is no standard requirement for the use of the "ergonomics" label, and it has been applied as much to a look as to the result of actual human-factors research. Toothbrush design has been a particular focus of this; over the last decade, a great many toothbrushes with striking new designs have appeared on the market **(figure 96)**, and there has been much confusion between those that reflect the thought and research required for the production of ergonomic toothbrushes, those that use the ergonomic look as a marketing ploy, and those created for style alone. In 1993, Philippe Starck set a design precedent with his much-celebrated toothbrush that borrowed its shape from the fluid, airy form of Brancusi's famous abstract sculpture of a bird in flight and came with its own display stand. Starck, taking his unique approach of situational thinking, understood the toothbrush as "a basic nonproduct. Everyone needs one, and they're all much the same. But," he

thought, "if, first thing in the morning, you had something bright and cheerful sitting on the shelf waiting for you, it would be like opening the bathroom window onto a summer landscape every day."[44] At the other end of the spectrum, Oral-B relied on extensive ergonomic and materials research for its CrossAction toothbrush **(figure 97)**, introduced in 1999,[45] with thorough documentation that is publicly accessible. During the studies, it was discovered, for example, that we use five different ways of gripping the handle of our toothbrushes; these were then thoroughly considered when the soft, polyurethane-foam handle was designed.[46] But consumers have no way of knowing which companies actually test their designs, as Oral-B did, and which merely design products to have what has come to be considered an ergonomic look. In the same way, the pen and razor industries have adopted an ergonomic vocabulary often without acknowledgment of whether serious testing has taken place.

Many products and systems designed specifically for, and often in collaboration with, those with disabilities also have found universal applications, especially in the area of communications. Text-to-speech software, created originally for the blind, has become widely used among the general population for retrieving e-mail messages over the telephone while closed captions, which make television accessible to those with hearing loss, are introduced in public areas where sound transmission is not viable or not desirable. Computers can now be readily adapted for the blind, for the color-blind, and for those with impaired vision, and many of those special features have substantial wider applications, too. Both Windows and Mac software have numerous accessibility options built into their operating systems, including voice alerts to let us

know about technical malfunctions, easy-to-effect type zooms, and the ability to change both the contrast and the color of the screen. Those with physical impairments who type with one finger or a mouth wand can use the sticky-key features that allow command keys and letter keys to be typed sequentially instead of having to hold two or three down at the same time.

The aging population, who gradually need more and more adaptive devices, and the disabled are served by an increasing number of purpose-designed objects and interior accommodations **(figure 98)**, which help them maintain independent and active lifestyles. Products as simple as plastic loops that attach to keys as an aid in opening locks and grips that convert doorknobs to levers, and as technologically sophisticated and complex as the Independence IBOT wheelchair, which Dean Kamen has been developing, exemplify their range. Kamen's amazing mobility-enhancing device will allow the chair to traverse rough terrain, mount a staircase, and raise itself on one set of wheels, making users feel as if they were standing up. But products designed for special needs can be not only physically enabling, they can be enriching and poetic, too. Blind, a yellow upholstery fabric from the Swedish group Saldo **(figure 99)**, is printed with what to the sighted may look like randomly patterned white polka dots but is recognized by blind people who touch it as Braille. The raised text aims to convey the sensation and emotional component of the color yellow in the words of six different authors. This fabric attempts to bring accessibility to a new level of inclusion by imbuing the design with a special message, although the young designers who created it are aware that their pursuit is not without controversy. "How do you describe something that you perceive with one of your senses to a person who doesn't have that

98
Kohler designed
several bathroom
and shower compo-
nents to satisfy the
needs of the aging
population and those
with strength or
movement impair-
ments, which meet
ADA requirements
(1998). But they also
created them with
universal design prin-
ciples in mind so that
they would not seem
special or be out of
place in any home.
(Courtesy Kohler Co.)

99
Blind, a fabric
designed by the
Swedish group Saldo,
aspires to give those
who are blind insight
into a realm that is
difficult to convey.
The Braille texts that
can be read on this
cushion, written by
several different
authors, describe the
sensation of the color
yellow for those who
have never been able
to see it.

particular sense?" Saldo asks. "Is it possible to give a blind person the 'sight' or 'feel' of a color even if that person has never seen yellow, orange, blue...? Can something beyond description be conveyed? Can you talk about silence or is that something that only fools do? How do you explain the color yellow to a person who is blind?"[47]

All of us, regardless of our physical needs, respond to products that satisfy our inner needs, that contribute to our comfort, spirit, and sense of well-being. We all feel better when a product meets our particular specifications, whether for fit, function, or style, and the large-scale availability of customization is now helping us achieve this. And we are enlarged emotionally when products evoke feelings within us and when we respond to the varied messages they convey.

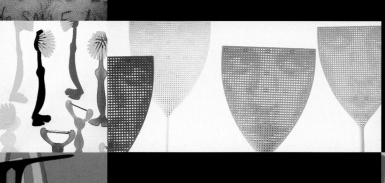

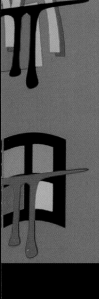

Messages

All designs tell us something. Their messages can appear openly in the form of text, image, or symbol, or they may be hidden, the subtle, layered communications conveyed through gesture, style, metaphor, identity, or branding. Even what may seem to be direct graphic messages may be laced with relativism, ambiguity, and irony, reducing the objectivity and clarity that had been expected from design in the past to a position where it can no longer be taken for granted. Stefan Sagmeister's poster advertising an upcoming lecture with text scratched into his torso **(figure 100)** is more than just the masochistic assault on his own body that it appears to be. It was meant as a wake-up call for graphic designers, who more and more had been relying on the handy tools of technology to achieve what seemed to be an unfeeling slickness and uniformity in their work. By broadcasting a mundane communication in this radical way, Sagmeister put himself on the line; he obliged his audience to empathize with his plight as they deciphered his very physical message. By so vividly rejecting the facility of the computer in order to find his own independent, tactile means of communication, he challenged his colleagues to take steps to discover their own paths of expression. Even symbols can be multidimensional, depending on context. We witnessed a remarkable reclamation and redefinition of a national symbol when, after the September 11, 2001, attacks, each of the expressive incarnations of the American flag **(figure 101)** conveyed a message of sympathy and solidarity that was immediately understood, creating a new meaning for a banal emblem and restoring a potency that had been lost long ago.

Products too tell us more than is immediately apparent. We saw how Nokia's interchangeable faceplates **(see figure 52)** and iMac's flavorful colors **(see figure 47)** reveal how we define ourselves by acquiring more and more products and then

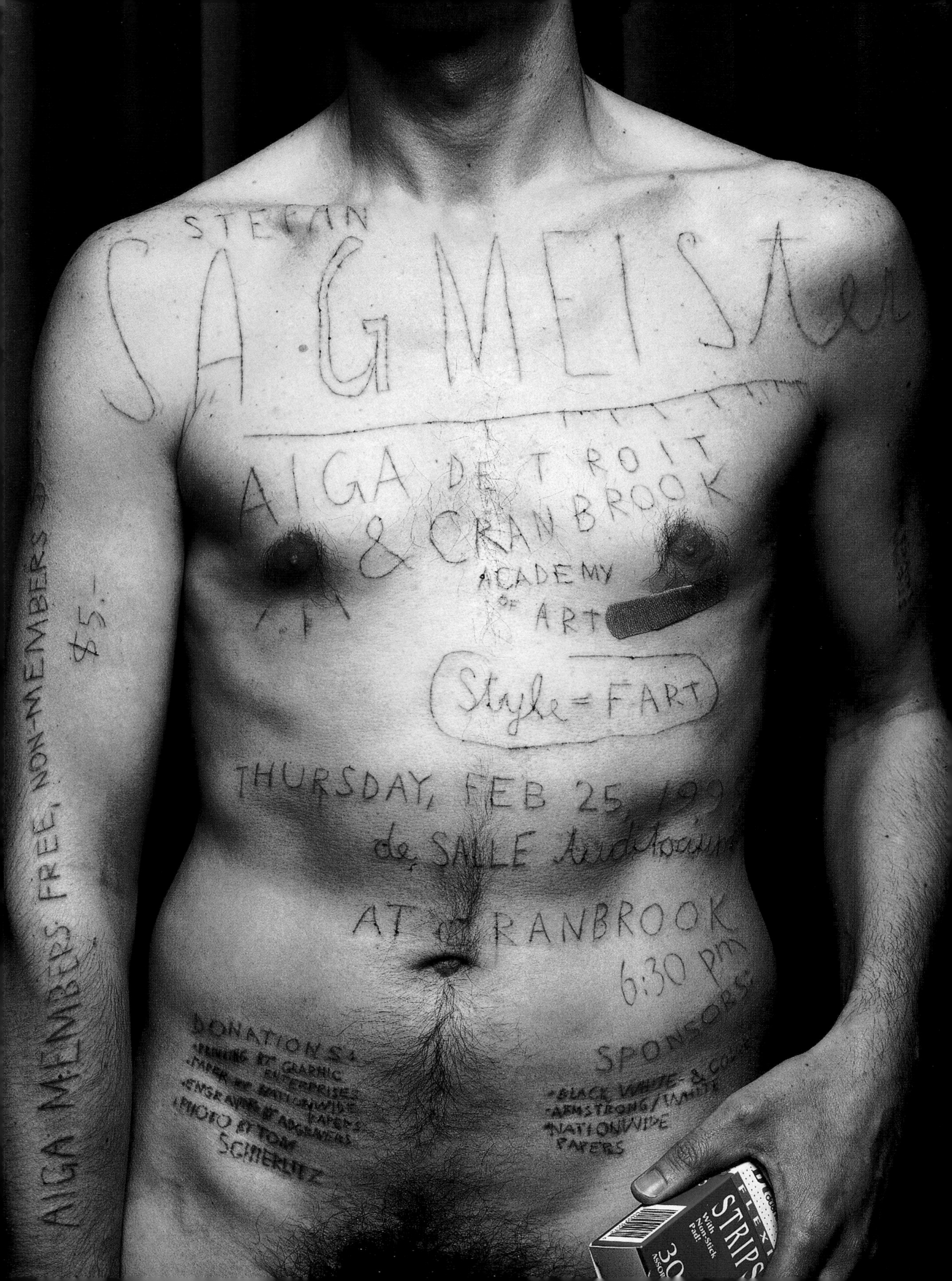

STEFAN
SAGMEISTER
AIGA DETROIT
& CRANBROOK
ACADEMY
OF ART
Style = FART
THURSDAY, FEB 25
de SALLE
AT
RANBROOK
6:30 pm
DONATIONS:
ENTERPRISES
PAPERS
SPONSORS:
•BLACK
•ARMSTRONG/
•NATIONWIDE
PAPERS
AIGA MEMBERS: FREE, NON-MEMBERS $5.-
STRIPS

100
In this poster for a lecture sponsored by the Detroit chapter of the American Institute of Graphic Arts (1999), Stefan Sagmeister's body conveys an implied message calling for the return of the designer's touch at a time when his profession had become slick and impersonal through its reliance on digital technology.

An assistant spent eight hours incising the lecture announcement on his torso. Notice how Sagmeister mocks the reality of that physical toll by holding a meager box of adhesive bandages.

101
After the September 11, 2001, attacks, many looked to their own domains to find vocabularies that would best express their personal feelings and reactions. The Weatherproof Garment Company used its own products to create this telling image of the American flag, which without any text at all could remind us that people from all areas and all endeavors shared in the tragedy.

separate ourselves from one another by customizing them. In accepting products as signifiers of status, we literally put them on pedestals as we display our sneakers **(see figure 23)**, our cameras **(see figure 41)**, our toothbrushes, and even our flyswatters **(figure 103)**. Now, with the arrival of hordes of everyday objects in figurative and narrative shapes, we are developing one-on-one relationships with what in the past were simply abstract and impersonal tools, revealing a need to personalize our environments. These anthropomorphizing products draw us in by the surprise of recognition or, more often, by simple whimsy. Koziol has produced an entire kingdom of creatures that inhabit its line of plastic "cutensils" **(figure 102)**, and Guido Venturini has created similarly colorful referential forms in his line for Alessi. Philippe Starck, too, inserts many such references into his household products, often, however, with ironic intent, and the reproachful faces that stare at us from his perforated plastic Dr. Scud flyswatters **(figure 103)** show just how ambiguous such product messages can be.

We live amid a profusion of imagery, and it is so fluid and so frequently manipulated that we cannot be certain that what we see depicted before us reflects the reality of our world. Today obfuscation is deliberately introduced —twists and turns provoked by designers that challenge us to experience the depths of meaning that may be found in any image and any message. These new approaches succeeded in the 1990s, especially among those who worked for youth-oriented markets, designing magazines of popular culture, compact discs, and advertisements for sports equipment such as skateboards and snowboards. Neville Brody in England and David Carson in the United States had early celebrity in this market; rather than following what had been the

102
Koziol's ever-growing
line of colorful, anthro-
pomorphic, plastic
kitchen and bath uten-
sils brings humorous
figurative imagery
into the field of
mass-market design.
With each object
given its own name,
we are asked to
empathize with these
cartoonish characters,
and they have become
popular for both their
form (as collectibles)
and their function.

103
The faces that
grimly stare out at
us from Philippe
Starck's Dr. Skud
flyswatters for
Alessi (1998) show
the dark side of
anthropomorphic
product imagery.

accepted rules of clean-cut layout and typography in commercial design, they introduced ambiguity and layering, using typography as expressive form itself. The design of mass-circulation magazines, advertising, and packaging soon followed their lead. By the end of the decade, the modernist dictums of the past had largely been discredited for a greater interest and more intense engagement on the part of the reader; alongside corporate logos and symbols (which themselves are not always the epitome of legibility), we are hit with a barrage of cryptic or skewed narratives in advertisements and television commercials that are no less ferocious for being implicit and subtle. At the same time that we have become more comfortable with these layered approaches to communication, the rumblings of a reaction against them have emerged with the increased status given to graphic designers who continue to follow, or reinterpret, the modernist outlook. Such large, international design firms as Pentagram and MetaDesign are especially effective in bridging the gap between the two approaches in their identity and information systems and graphics for large institutions and corporations.

The postmodern ideal of engagement has entered the world of typography as well. Software programs such as Fontographer have enabled newcomers as well as professional type designers to turn their own computers into type shops and to do in hours what type designers and type houses in the past had taken months or years, and large expenditures, to do. With this easy access, tens of thousands of new, homegrown typefaces have been issued, many by upstart companies. Some designers, such as Matthew Carter in the United States and Jean-François Porchez in France, rethink and update earlier typefaces while others go off in new directions, demonstrating little fealty to

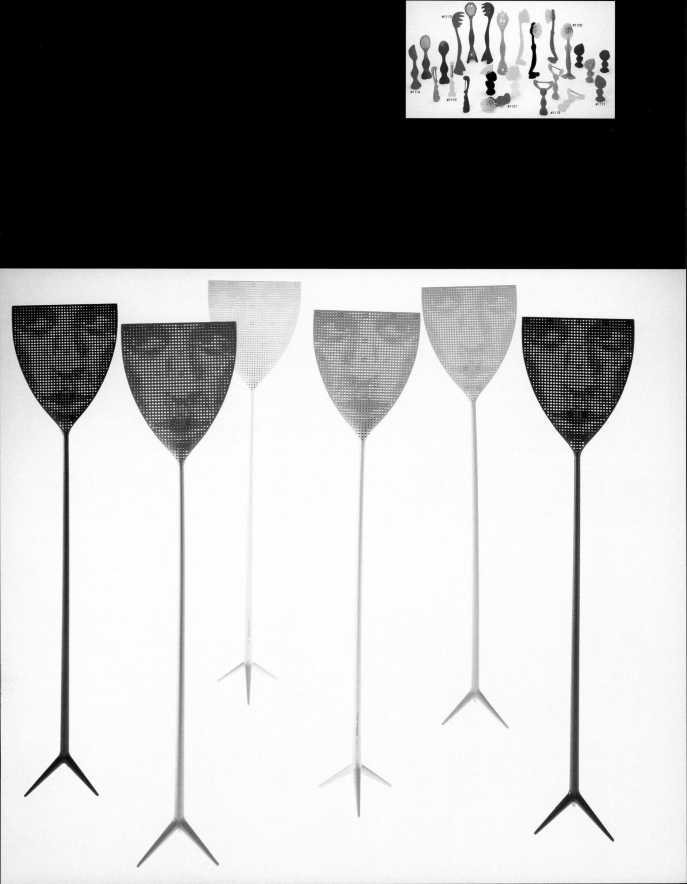

AA BB BCCCC DD
EEE FFF GGGH H
III JJJ KKK LLLL
MINNOOOP PQQRR
SSS TU UVVVW W
XXXYYYZ !@£$%&

11 22 33 44
66 77 88 999 0
#¢ .;:"?/,+$$ⒶⒸⓇ ¥

AA IIOO USSKMM
11 22 33 444
6 67 78 888 999
#¢ .;:"?/ $£ⒶⒸⓇ ¥

Mulinex Regular
Cavolfiore
Coltello Regular
Forchetta Regular

Ian Swift's Gunshot 6 Bore type, seemingly riddled with bullets, adds its own meaning to the meaning of the typographic message it is used to convey.

Alessio Leonardi sees his type as "actors" in the drama of communication. His Alessio faces for FontFont, which include Mulinex, Cavalfiore ("cauli-flower"), Coltello ("knife"), and Forchetta ("fork"), are, he says, "the fruit of rich dreams, born from a life lacking sleep and food. Perhaps I should say 'nightmares' when I look at the results. The names of the faces remind you of kitchen utensils and electric household appliances that don't work, knives that are blunt, and over-cooked cauliflower needing salt."

the strictures of historic typography and the modernist preference for structure, clarity, and legibility. Faces such as Ian Swift's Gunshot 6 Bore, in which the alphabet looks as if it had been fired at with buckshot **(figure 104)**, demonstrate a narrative approach that brings wit and irony to type today. Alessio Leonardi, an Italian designer, parallels this in his work, considering his letter forms as "actors" **(figure 105)**: "I have tried to transform alphabetic characters into pictures," he says, "precisely the opposite of the development of the alphabet, which was created from ideograms. The signs and drawings...tell the story contained in the text, but at the same time transport it to a further narrative level; it is its own story, completely accidental, not explicit, but open to interpretation."[48]

A narrative, associational viewpoint also has invaded the vocabulary of icons. Until the last decade, objective form and immediate recognizability were the undisputed requirements for information and wayfinding graphics, especially in a shrinking world where people from different cultures who speak different languages must be accommodated. This was best demonstrated in the familiar universal travel symbols that we see at every airport and on every restroom door, created for the U.S. Department of Transportation by Cook and Shanosky under the auspices of the American Institute of Graphic Arts between 1974 and 1979. Now alternative approaches challenge this dogma. Throughout the 1990s, in such highly visible venues as the Olympics, for example, the richness of content shared design decisions with clarity of message as function extended beyond wayfinding to include associational overtones. For Lillehammer in 1994, Sarah Rosenbaum designed mysterious and difficult-to-recognize pictograms

106
Malcolm Grear
Designers looked
to the ancient origins
of the Olympics as
inspiration for the
referential pictograms
they designed for
the Atlanta summer
games in 1996. Their
naturalistic rendering
of the symbols for
each of the sports,
based on the figures
of ancient Greek vase
painting, challenged
the dependency on
impersonal geometry
that had been the
hallmark of Olympic
pictograms in the
past.

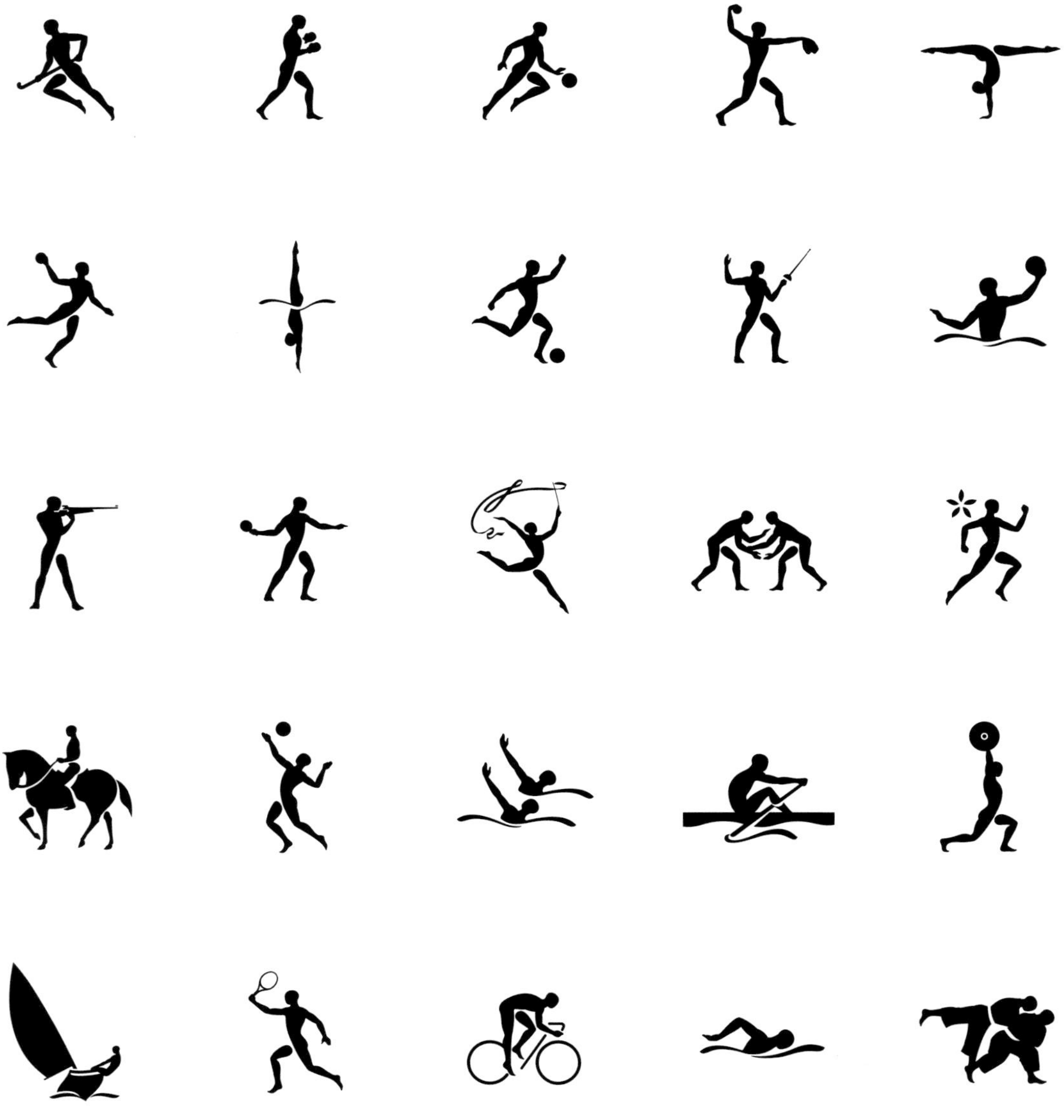

identifying the winter sports that were based on the jagged forms of ancient
Norwegian rock-cut imagery, giving them a nationalistic slant. For the summer
games in Atlanta in 1996, the symbols created by Malcolm Grear Designers
alluded to the painted figures on ancient Greek vases and thus the origins of
the Olympic movement (figure 106); svelte and naturalistic, they were applauded
for their historic associations, their legibility, and their elegance as they
showed that geometry was not the only possibility for information graphics.
At Salt Lake City in the winter of 2002, the angular, almost cartoonish, forms of
the pictograms created by the venue's design team (figure 107) were based on the
branding irons that introduced an early identity system to the Old West.

Today's idea of branding is much more complex than simple identity. It seeks
to establish lifestyle concepts for products, expecting customers to associate
themselves closely with brands as mirrors of their own experiences and values.
Branding includes all aspects of the public image of a product or service, from
its form and function to its name and logo, its advertising and marketing
program, which Martha Stewart has exploited so successfully in her television
series, publications, and product lines (figure 108). Establishing a brand identity
takes time, the most successful relying opportunely on a continuing campaign
that keeps interest alive. This is the rationale for the long-standing barrage
of Absolut images from the Swedish distiller, where we find the form
of its vodka bottle wrenched out of all types of landscapes and represented
by artists in all types of materials, and the unspoken, romantic narratives
implied by the protagonists of Ralph Lauren's stunningly photographed
clothing advertisements.

Everything from
mothballs to muffin
tins is similarly iden-
tified in the Martha
Stewart Everyday line
of products for K-Mart
to reinforce the brand-
ing experience that
Stewart epitomizes.
It features both updat-
ed color-coordinated
versions of vernacular
objects from the past,
such as this pitcher
and mugs, and origi-
nal, utilitarian
designs.

Some brands extend their lines to supporting products, as Harley-Davidson has done with its move into clothing accessories, while others have pushed into new areas, depending on the celebrity of their names to infiltrate new markets successfully. Still others seek affinity partners to help expand their own realms, as Adidas did when it invited designers from disparate disciplines to refresh the realm of sport-shoe design. Audi and Adidas, two German firms, teamed up to create the Kobe shoe, with Audi bringing automotive-performance know-how to the shoe's construction, meant to reflect the laid-back sophistication of the Lakers' star player Kobe Bryant. Adidas also linked up with the hip Japanese designer Yohji Yamamoto to create sport shoes for women with a fashionable edge.

Brands now seem limitless in their reach; we recognize thousands of them and we are continuously bombarded by their names, their logos, and their advertising. We often rely on the messages of brand identity to guide us when we shop for products and services, and most of us at one time or another have been lured on by the signature of a famous designer or the logo of a big-name manufacturer, much as we may have tried to separate our own desires from those that branding has taught us to revere **(figure 109)**. In a similar but simpler way, we convey our own messages and create our own branding when we choose products for ourselves. Through the clothes we wear, the products we buy, and the way in which we live, we create a personal style, or identity, and in doing so reveal aspects of what kind of people we are and what kinds of values we have.

Afterword: Making Choices

How might an awareness of the issues of design affect us when we go shopping? For most of us, style, price, and functional features are our basic considerations, but even if we want to learn more about other aspects of our products, we have little to go on. Why shouldn't we have more information before we hand over our credit cards? What would happen if all manufactured goods were labeled as completely as packaged foods? The American Automobile Labeling Act required this to a certain degree, and we are supplied with data on gasoline consumption and the percentage of American components in our cars and even told where they were assembled. Should not the concept of traceability be applied to other products as well? Appliance labels list their energy efficiency, but what about letting us know where all the components come from, their ability to be recycled, and their toxicity or sustainability? Why shouldn't we know the name of the designer and the manufacturer, how the product ranked for safety and accessibility, its environmental impact, the origins of its materials, the energy expended in its manufacture, and how those who labored on it were treated and compensated? But ask as we may, answers will be few, since little of this information is made available.

Not all buyers read food labels closely and not all consumers would read such product labels, but the information should be there for those who want it. But regardless of what the label told us, if we loved the look of an object, we might very well buy it anyway, because we gain pleasure and satisfaction by surrounding ourselves with things that we feel are beautiful. How an object appears creates a first impression, sending a message that we cannot overlook. Either it speaks to us and we want to buy it or it leaves us cold, more suited to those with other tastes. Thinking about design today means thinking beyond aesthetics and style, however. It means recognizing the complex issues of production, technology, responsibility, and individuality and the messages that underlie our products and their branding. In thinking about design, we confront the divergent values that we find in society today. As times change, design changes also, and some of the attitudes and trends that typified an era of optimism and consumption also have begun to change. Where design will go remains to be seen, and while many of the same questions we have asked about design today will be asked again tomorrow, we have no idea whether the answers will be the same or completely different.

Notes

1
Anthony Bertram, *Design* (Harmondsworth, England: Penguin Books, 1938), 11.

2
Edgar Kaufmann, Jr., *What Is Modern Design?* (New York: The Museum of Modern Art, 1950), 7.

3
See Robert Blaich with Janet Blaich, *Product Design and Corporate Strategy: Managing the Connection for Competitive Advantage* (New York: McGraw-Hill, 1993).

4
This discussion is dependent upon the presentation of process in the exhibition and accompanying brochure *Design @ Work: The Process Behind Products,* curated by Mary Douglas and shown at the Mint Museum of Craft + Design in Charlotte, N.C., from July 7 to October 7, 2001.

5
See Paul J. Smith, ed., *Objects for Use: Handmade by Design* (New York: Harry N. Abrams, Inc., Publishers, in association with the American Craft Museum, 2001).

6
Malcolm McCullough, *Abstracting Craft: The Practiced Digital Hand* (Cambridge, Mass., and London: MIT Press, 1996).

7
Stanley Lechtzin, "Digital Divergence... Towards a Digital Craft," unpublished paper presented at the Virtual Revolution Symposium at the American Craft Museum, New York, November 2, 2001.

8
Steelcase Design Partnership, New York, *Gaetano Pesce: Modern Times Again* (November 10 through December 8, 1988), exhibition brochure, n.p.

9
Alvin Toffler, *The Third Wave* (New York: William Morrow, 1980), 195–209.

10
See Joseph Pine, Jr., *Mass Customization: The New Frontier in Business Competition* (Boston: Harvard Business School Press, 1999).

11
Toffler, cited above, 201.

12
Howard Ketcham, "Colors Car Buyers Want," *American Fabrics,* no. 28 (Spring 1954), 113–14.

13
<http://www.us.levi.com>

14
Made To Order Online Q&A; <http://nikebiz.com/media/mto_qa.shtml>

15
<http://www.dell.com>

16
<http://www.lindal.com/home.cfm>

17
<http://www.alaskasoapmill.com>

18
See Naomi Klein, *No Logo: Taking Aim at the Brand Bullies* (New York: Picador, 1999); see also Steven Flusty, "Icons in the Stream: On Local Revisions of Global Stuff," in Aaron Betsky, *Icons: Magnets of Meaning* (San Francisco: San Francisco Museum of Modern Art, 1997), 52–65.

19
See, for example, Ad-Hoc Committee on Sweatshop Labor, "A Code of Workplace Conduct for Penn Apparel Licensees," *Almanac* (University of Pennsylvania), March 28, 2000.

20
See <http://www.cepaa.org>

21
Jasper Morrison, comp., *A World Without Words* (Baden, Germany: Lars Muller, 1999).

22
Tom Dixon, *Rethink* (London: Conran Octopus, 2000), 60.

23
See Joan Kron and Suzanne Slesin, *High Tech: The Industrial Style and Source Book for the Home* (New York: Clarkson N. Potter, 1978).

24
Karim Rashid, "Blobism," in *Karim Rashid: I Want to Change the World* (New York: Universe, 2001), 245.

25
Delphine Hirasuna, "Sorry, No Beige" [interview with Jonathan Ive]; <http://www.apple.com/creative/collateral/ama/0102/imac.html>

26
Quoted in "O'Neal Signs Deal Worth $88.5 Million," *New York Times,* October 14, 2000.

27
Ray Kurzweil, *The Age of Spiritual Machines: When Computers Exceed Human Intelligence* (New York: Viking, 1999).

28
Marshall McLuhan, *Understanding Media: The Extensions of Man* (New York: New American Library, 1964).

29
Marshall McLuhan and Bruce R. Powers, *The Global Village: Transformations in World Life and Media in the 21st Century* (New York and Oxford: Oxford University Press, 1989), 80.

30
McLuhan, *Understanding Media,* cited above, 19.

31
<http://www.w3.org/People/Berners-Lee>; see also Tim Berners-Lee with Mark Fischetti, *Weaving the Web: The Original Design and Ultimate Destiny of the World Wide Web by Its Inventor* (New York: HarperCollins, 1999).

32
David Siegel, *Creating Killer Web Sites: The Art of Third-Generation Site Design,* 2nd ed. (Indianapolis: Hayden Books, 1997), 4.

33
Arik Levy, "Designer's Talk–Infinite Light," <http://www.snowcrash.se>

34
Ezio Manzini, *The Material of Invention* (Cambridge, Mass.: MIT Press, 1989), 29.

35
See, particularly, the report of the testimony of John Ahmann, a developer of the VotaMatic machine for IBM, where he acknowledged the problems of the machine and that he had earlier attempted to improve it; David Barstow and Dexter Filkins, "For the Gore Team, a Moment of High Drama," *New York Times,* December 4, 2000.

36
Keith Bradsher, "Close Study of Explorer's Design Reveals a Series of Compromises," *New York Times,* December 7, 2000.

37
Donald A. Norman, Preface, *The Design of Everyday Things* (New York: Doubleday, 1989), p. x.

38
Martand Singh, "The Precious Object: The Role of the Hand-Crafted Product," in Alberto Cannetta and Soumitri Varadarajan, eds., *Quality Through Industrial Design* (New Delhi: Mosaic Books, 1992), 126–27.

39
Yemm & Hart, Marquand, Mo., "Origins: A Recycled Plastic Material," product information sheet, rev. June 4, 2001.

40
Andrea Truppin, "William McDonough: 1999 Designer of the Year," *Interiors,* vol, 158, (January 1999), 95–117; Roger Rosenblatt, "Heroes for the Planet: William McDonough, The Man Who Wants Buildings to Love Kids," *Time,* February 22, 1999, 70–73.

41
"The McDonough Braungart Sustainable Design Protocol"; <http://www.mbdc.com>

42
See Lars Muller, ed., *Freitag: Individual Recycled Freeway Bags* (Baden, Germany: Lars Muller, 2001).

43
See Niels Differient, et al., *Humanscale: A Portfolio of Information* (Cambridge, Mass.: MIT Press, 1974, 1981).

44
Quoted in Conway Lloyd Morgan, *Starck* (New York: Universe, 1999), 11.

45
Information about the Oral-B CrossAction toothbrush was presented in an unpublished undergraduate paper by Beth Bowers in my course on Twentieth-Century Design at the University of Pennsylvania, Philadelphia, in 1999.

This book was conceived and written independently, but it owes much to the many perceptive suggestions of Hilary Jay, director of the Design Center at Philadelphia University and co-curator of the traveling exhibition it inspired. Her reactions to various stages of the manuscript and her familiarity with the world of design helped me broaden my coverage and hone many of the points that I had wanted to make. Ralph Lieberman gave his customary valuable advice and support, and my two editors at Abrams, Eric Himmel and Richard Olsen, responded with decidedly personal and divergent viewpoints, challenging my take on the subject as they offered their own studied opinions of design today. John Klotnia at Opto Design provided a most sympathetic format for conveying the message of this book, aided by Will Brown's special photography. Constantin Boym, Laurene Boym, Mike Flanagan, Guido Könn, Dan Lazarchik, Stanley Lechtzin, Josh Owen, and Marcel Wanders also generously aided this undertaking.

My knowledge and understanding were greatly enhanced by the contributions of the students in my undergraduate courses on the history of design at the University of Pennsylvania, who have kept me in touch with the meaning of design in the lives of the next generation. Their research papers devoted to specific products brought new objects to my attention, gave me insight into other outlooks, and added to the richness of my thinking. Many products discussed here were the subjects of their guided research papers, and I wish to acknowledge any contributions that their discoveries and ideas may have added to this book: Apple iMac (Oliver Benn, Angela Getz, Kathleen McDonough), Bandai Tamagotchi (Andrea Bertoline), Casio wristwatch (Kelly Lwu), Chrysler PT Cruiser (Debra English), Gillette Venus razor (Sarah Aibel, Rebecca Marshall), Handspring Visor/Palm Pilot (Jane Branton, Joshua Drazen, Joshua Skaroff), IKEA products (Sean Gannon), Koziol utensils (Jennifer MacDonald, Jennye Stubblefield), Nike sport shoes (Jarred Ballou, Michael Winston), Nokia cellular telephone (Evan Goldberg, Beth Tobias, Stephen Ward), Oral-B CrossAction toothbrush (Beth Bowers), OXO products (Katherine Fraser, Christy Gressman), Sony PlayStation (Jeffrey Doyon), Target's Michael Graves products (Benjamin Langsfeld, Scott Rigby), TDK CD holder (Claire Pinto), Umbra's Karim Rashid trash can (Manuel Rabate), and Volkswagen Beetle (Anne Hankey).

Issues

Busch, Akiko. *Design Is*. New York: Princeton Architectural Press, 2002.

Hawken, Paul, Amory Lovins, and L. Hunter Lovins. *Natural Capitalism: Creating the Next Industrial Revolution*. Boston: Little, Brown and Company, 1999.

Julier, Guy. *The Culture of Design*. London: Sage Publications, 2000.

McDonough, William, and Michael Braungart. *Cradle to Cradle: Remaking the Way We Make Things*. New York: North Point Press, 2002.

Norman, Donald A. *The Invisible Computer: Why Good Products Can Fail, the Personal Computer Is So Complex, and Information Appliances Are the Solution*. Cambridge, Mass., and London: MIT Press, 1998.

Papanek, Victor. *The Green Imperative: Natural Design for the Real World*. New York and London: Thames and Hudson, 1995.

Products

Albrecht, Donald, et al. *Design Culture Now: National Design Triennial*. New York: Cooper-Hewitt, National Design Museum, Smithsonian Institution, and Princeton Architectural Press, 2000.

Branczyk, Alexander, et al., eds. *Emotional Digital: A Sourcebook of Contemporary Typographics*. New York: Thames and Hudson, 1999.

Byars, Mel. *On/Off: New Electronic Products*. London: Laurence King Publishing, 2001.

Castelli, Clino T. *Transitive Design: A Design Language for the Zeroes*. Milan: Electa, 1999.

Datschefski, Edwin. *The Total Beauty of Sustainable Products*. Crans-Près-Céligny, Switzerland: RotoVision, 2001.

Fiell, Charlotte, and Peter Fiell. *Designing the 21st Century*. Cologne: Taschen, 2001.

Industrial Designers Society of America, ed. *Design Secrets: Products: 50 Real-Life Projects Uncovered*. Gloucester, Mass.: Rockport Publishers, 2001.

Leibrock, Cynthia A., and James Evan Terry. *Beautiful Universal Design: A Visual Guide*. New York: John Wiley and Sons, 1999.

Lupton, Ellen. *Mixing Messages: Graphic Design in Contemporary Culture*. New York: Cooper-Hewitt, National Design Museum, Smithsonian Institution, and Princeton Architectural Press, 1996.

Lupton, Ellen, et al. *Skin: Surface, Substance, and Design*. New York: Cooper-Hewitt, National Design Museum, Smithsonian Institution, and Princeton Architectural Press, 2002.

Redhead, David. *Products of Our Time*. Basel: Birkhäuser, 2000.

Smith, Paul J., ed. *Objects for Use: Handmade by Design*. New York: Harry N. Abrams, Inc., Publishers, in association with the American Craft Museum, 2001.

Williams, Gareth. *Branded?* London: V&A Publications, 2000.

Index